HOW TIME DILATION CREATES
PROPULSIVE GRAVITY

HOW TIME DILATION CREATES
PROPULSIVE GRAVITY

Gravity from Energy Potential with Unification of Forces

By

Scott Calvert

HOW TIME DILATION CREATES PROPULSIVE GRAVITY

Gravity from Energy Potential with Unification of Forces

Also by Scott Calvert

Table of Contents

Table of Illustrations

Preface

The topic of gravity has been on my mind and occupying my thoughts for the better part of my entire life. The reason for that is in my mind there has never been an adequate explanation that has described it with complete detail. The commonly accepted theory for gravity leaves a lot to the imagination which essentially means there are gaps and missing links between pieces of the puzzle.

My mind functions analytically and those gaps don't compute. Even up to this day the information regarding gravity usually ends up with disclaimers such as "we don't know this or that" but eventually those questions will be answered. In my opinion they won't be because if we follow the same path that created those questions then we are moving further away from the answers.

I have spent many hours contemplating on the process that creates gravity and of course have had my missteps.

When one idea hits a dead end for me it has always transitioned to a more promising line of reasoning and expansion of the process. The common denominator is that I have never stopped trying to work it out and dead ends were just transition points to a higher level. When I had worked out the previous version of gravity I felt the need to share it and get it out into the public well of knowledge. That was the genesis of the first edition of this book.

Unknowingly I didn't realize that the process could be morphed into a wider scope and graduate to a higher level. It is also more scientifically acceptable with each step of the mechanism being acceptable on its own with the sum of its parts being greater than the whole. I believe now that the process of the mechanism for gravity I have developed is finally complete. That there is no other direction it can go and explains gravity to the letter. It is the end of working out the process in my mind, not a dead end but a completion with the journey reaching its destination.

This book is actually a second edition but with a different title. The original title was "How Time Dilation Creates Quantum Gravity". I could not continue to use the original title because it denoted a description of gravity as quantum which has a very specific meaning. The meaning being one of describing gravity as it regards to quantum mechanics. The word quantum relates to quantizing gravity and alludes to proposing how it functions in relation to the existence of a graviton particle. I do not believe in gravitons and my explanation of gravity does not include them. My view and the description I use changed between the two versions of this book. Because of that I could no longer represent this book using the old title.

The original premise of the book was explaining gravity as it relates to the subatomic level between the electron and the nucleus with gravity the result of a higher relative electron speed in the faster area of time. It is now expanded to a point where gravity is manifested in the range where anything of mass exists. The reason is that the characteristic of mass is the result of being a field. This is for anything where its energy has opposing constituents that create a field. Gravity is simply a potential created in that field by differences of space-time as a result of diminishing time dilation. With the genesis of time dilation manifesting because of a governing action that space-time has on energy to limit its speed.

After the first edition of this book was published I began writing a manuscript for a peer reviewed journal. At the time I thought I had finished with the theory but during that writing process over a few months secondary pieces began to take precedence and became primary. The original theory began to grow and expanded exponentially to a higher level over that which was laid out in the first edition. Because of that I must add the new material and create a second edition as well as retitle it in order to properly represent the content. I don't consider it a mistake but simply being unaware that the journey was incomplete.

In the previous iteration I put the emphasis of the mechanism of gravity on the difference of relative speed of electrons as they orbit the nucleus in an atom. I left the spatial difference due to time dilation open to interpretation because I didn't know of any specific effect that may have and thought it was just a side effect of the relative space-time created by time dilation. The spatial difference caused by time dilation actually creates for atoms energy potential in its fields that shows a much more precise and scientifically acceptable reason for the mechanism of gravity. The two factors of relative electron speed and relative spatial dimension should both be a potential cause for gravity but the later has a broader and more stable foundation which also makes it easier to connect the mechanism of gravity to the process that induces force.

The direction of this book is going to follow the same path and create a theory of gravity not previously considered. This is an expanded version of the first book with a higher level reasoning as to why and how time dilation creates gravity by reason of relativity, spatial differential and energy potential. There is also an aspect of electromagnetism and the strong force which creates gravity since gravity manifests in any field that is considered to create the characteristic of mass. There may also be some philosophical changes to my opinion as to how space and time or space-time relate to each other. I will hold on to their inverse relationship and elaborate a little more on it.

It is an interesting side effect that this updated version has electromagnetism and the strong force as factors that initiate the mechanism of gravity. That fact alone should emphasis the importance of the new material. That gravity is connected to fields produced by electromagnetism and the strong force placing gravity under their purview and essentially combining gravity with two of the fundamental forces of nature.

Preface

Einstein spent 30 years trying to unite general relativity (GR) with electromagnetism unsuccessfully but had he realized that meant GR was incomplete he could have revised it accordingly. I am simply following in the footsteps of Einstein trying to work out the precise mechanism of gravity. It just so happens that by defining the relationship between space and time more precisely it shows another avenue GR could have taken.

I will redefine gravity thus redefining general relativity to show gravity as a force and as no longer a separate fundamental force of nature. The content in this book finalizes Isaac Newton's work on gravity by explaining the mechanism of how gravity functions which he left up to the reader to conceive. This book also redefines Albert Einstein's general relativity by showing the precise mechanism of the relationship between time dilation and how that directly creates gravity. It creates gravity not as an attraction or "not a force" but as propulsion and propulsive force from a higher potential energy induced in an object of masses fields.

Introduction

Have you ever imagined what causes the earth to exert a gravitational attraction on the moon and the same in reverse? It is hard to comprehend just how that invisible action at a distance takes place from one object to another. It is an insurmountably task to try and figure out how one object of mass pulls on another. What is reaching out through space to grab the other object? How is it grabbing the other object, is it by electromagnetism or some sort of wave or particle? What is the medium that is transmitting these possible factors when there is seemingly nothing in between them?

"Seemingly nothing" is only a description but not the reality, what lies precisely between the earth and moon is space-time. To be more precise time dilation of varying degrees (relative space-time) lies between the earth and the moon and that is the underlying factor behind gravity. If your mind has been working out the details and you cannot link time dilation to the earth attracting the moon then you can stop pondering that thought. Time dilation and relative space-time does not create an attractive force of gravity it creates a potential energy in objects of mass. That energy induces a force which intern creates motion. It just so happens that the motion of an object is toward the direction of the greatest source of time dilation acting on it from any point around it. The greatest source creates the greatest potential energy which overpowers any other source. Simply put if an atom or object has multiple sources of gravity around it then the strongest time dilation will create the largest force of propulsion and motivate the object to move toward it.

Gravity is more of a natural means of propulsion than an attraction between two objects. The earth isn't pulling on the moon because really the moon is reacting to the time dilation of the earth which gives the moon the force to move toward the earth. The moon stays in orbit and doesn't crash into the earth because the orbital force equals the amount of force that gravity is inducing within the moon.

That is a basic summary of gravity but before we disassemble the exact process from the first to the last page of this book some ground work needs to be done. As a precursor we need to cover in more detail the topics of time dilation and the characteristics and relationship between space and time. You will see how clear and easy defining gravity really is and you will wonder how it could have taken this long to come to this conclusion. At least it's my conclusion, changing minds about this I'm sure will not come easily.

Introduction

***

The conception of gravity starts with time dilation but that in itself has its own conception. That process starts as space-time regulating and governing all energy and therefore all structures of energy as mass. The mechanism of gravity starts at the smallest structure of energy that comes together to form a field. Two known fields are of electromagnetism and the strong force which are fundamental forces. The largest structure of energy that forms a field is an atom. The smallest field is that of the protons and neutrons or nucleons which main field is created by gluons as a gluon field.

Any structure of energy in between those two that forms a field would add to the effect. Gravity would manifest in those fields and create a compounded effect of propulsion. An accumulation of atoms also compounds the effects. The action of gravity is specifically field energy potential related while energy without an opposing field structure does produce time dilation it lacks specific characteristics that create what we know as gravity. The effect of the regulatory action of space-time induces the force of gravity in objects of mass via time dilation whether they are part of the same mass or at a distance.

One of the main reasons gravity has had such an elusive understanding throughout history is because it has always been described as an attraction. Using that explanation or characteristic falsely identifies the main factor of the phenomenon thereby any conclusion or theory derived from it would be false. I had a pivotal moment in my quest to explain gravity which came after developing my theory when I realized gravity is not a force of attraction at all. It is in fact a force of propulsion naturally created in atoms and fields to propel them to the nearest and greatest source of time dilation. I use atoms as an example because they make understanding the process easier. A nucleus of an atom and a nucleon would also suffice due to its structure as a field.

Gravity has always been thought of as an attraction and the only other possibility would be a push together rather than a pull together. Since the theory of gravity has never really been accurate when describing it as a pull of matter (an attraction) then the only other explanation is that it's a pushing of matter towards other matter, a propulsion. The word gravitation seems to eliminate the assumption of either descriptions and up to this point was the best way to present it but that is an exception to the rule which most don't follow.

Objects of matter will gravitate toward one another by means of a continuous force that acts like an atomic level rocket engine that never runs out of fuel while experiencing relative space-time. Every single atom that exists in the universe has this potential engine while being powered by time dilation. It has the ability to become an engine and also have induced in it a difference of potential within its fields. When it is in a greater field of relative space-time it produces a greater force and lesser force in a lesser field. Zero force is created in "theoretical" flat space-time.

Another reason that gravity has been so difficult to pin down an explanation for is that there is no conceivable way an object reaches out and affects another to draw them closer together to attract them. In a quest to try to explain that the theories of gravitational waves and graviton particles came into being, I have heard the same information as all of you on what creates gravity, it is a wave, a particle or both or the curvature of space-time. I say it's none of those even though time dilation may be misconstrued as creating curved space-time and that creating an abstract view of gravity.

It wasn't until I realized that the only thing that spans the distance between two objects of mass is the relative space-time of time dilation and that's how I came to a final understanding of how gravity functions. I've had my ups and downs in my attempts to work out gravity in my mind, it did take time. The greater part of my existence on this planet but I finally worked out how the process of gravity works. You could call it a passion project of mine because I considered the topic incomplete which added to the fascination to work out the puzzle.

***

I had previous theories to try and explain gravity as attraction but they would always stop short of being able to fully explain how they functioned. I could not work out the mechanism of an atom pulling on space and the intersecting line of another atom pulling on space. There is no way to conceive of an atom having a direct ability to exchange a force in the medium of space. That aspect was a stretch and not good enough for me, if a theory doesn't explain in every detail or step how something works then there is a gap that cannot be spanned making the theory improbable. It's like having puzzle pieces missing and you have no way to link the pieces on each side of the gap. When I looked at things from a very simple perspective the reality of how gravity functions began to unveil its picture to me.

How Time Dilation Creates Propulsive Gravity

I think lack of simplicity is the biggest issue with most theories spanning the universe and that of gravity from general relativity to the current day, they are much more complicated and complex than they need to be. Theorists seem to think that the more something is unknown the more complex a solution must be. A journey of a thousand miles begins with a single step, start with step one and continue without jumping from beginning to end skipping vital links from one step to another.

The entire universe starts out as loose elemental particles and the governing action of space-time "time dilation" is the mechanism that creates everything in the universe and makes it what it is today. It starts the concentration of matter beginning with hydrogen and then gravity forms nebulas to create more complex atoms. It is also the main factor in the process of a regenerating universe or universes because as the process continues each universe births new universes. The universe does not have to end in order for a new one to be created. Multiple universes will exist in different iterations of progress simultaneously. The "in between" will simply evolve from birth to death like the generations of life.

In order to accomplish the task of defining gravitation as a simple process we will have to be specific about the relationship between space and time since they are proportional and inseparable. By using time dilation and the characteristics of space and time which are the true sources of gravity a clearer more defined picture of gravity will unfold. A gravity in its nature acting as a force since it clearly shows that in its characteristics and in our everyday experience of the effects of gravity.

Introduction

***

It took me about 25 years to come to my conclusions about space, time, gravity and the universe. When I look back at it I always enjoyed the process of contemplating the nature of things that in my mind we had no explanation for because the explanations we did have weren't good enough. I enjoyed the thought process and looked at it as a puzzle as well as therapy or a diversion. It involves putting your mind to work on processes much bigger than yourself. Many times as I fall asleep at night I would go over certain pieces of the puzzle in my mind to see how they functioned and fit together with the other pieces already conjured up in my thoughts. Instead of counting sheep I would imagine the effects of atoms on space to try and figure out how one atom could attract another. I would ponder not only about gravity but of the whole universe and how it functions as well as black-holes and what would be inside of them and more.

The first book on the subject of the universe was the beginning of the last stage of this journey. The year 2020 was about to start, I knew I was ready to start writing because I had finally worked out the last piece that was unknown to me or so I thought, the mechanism of gravity. I figured it out about 8 months prior and spent that time pondering it to see if that puzzle piece fit but as someone with little to no experience writing a book I didn't know how to put the words down and formulate them eloquently. After I got started I knew I had a lot more to learn about writing. So that's what I did next by absorbing all I could about writing and publishing a book. I learned what I could to get myself started it just took a little longer since I was at the beginning of the process. The process is like working out where repetition creates results, write, proof read, edit and repeat.

In the 25 years I had contemplated about and focused on the universe and every aspect in it, my first book was a big task. I covered all of my new theories and redefined the entire universe. It is written in a rational and common sense way to explain everything from black-holes, dark matter, and dark energy to gravity and last but not least the real actual way the universe is recreated and gets reborn. But since I began a journey I knew nothing about which was writing a book, a vague and general title did little to emphasize what was between the covers. I knew almost immediately or within about a month of publishing it that I needed to take a further step and that was to write another book to highlight the hidden gem that lay inside of my unknown book. I consider my theory of gravity that hidden gem since it's the first of its kind and I believe last in describing the action of gravity accurately. That is the genesis of how this book came to be, explaining that and my thought process may be helpful to you in learning a little more about the journey I took to get here.

***

After finishing and publishing the first edition of this particular book originally under a different title I endeavored to write an article to a peer reviewed journal. During that process of having to explain it with scientific writing along with time and focus I was able to elaborate more on the characteristic of space-time differential (relative space-time), the inverse relationship between space and time with the diminishing effect of time dilation. From those things bouncing around in my head for many years all I needed was to take one step further to get to the next level of understanding. Once there I was able to progress this theory of gravity to its last incarnation that leaves no stone unturned in explaining the mechanism for gravity.

The fine tuning of my thoughts on the inverse relationship between space and time allowed me to see how the action of gravity is directly related to fields such as with electromagnetism and the strong force and fall under their heading. That gravity isn't a separate fundamental force of nature and only has the manifestation of fields to thank for its realization. That is one of the reasons I was compelled to write a second edition of this book as the repercussions of that statement mean I've unified gravity with other fundamental forces. The characteristic that defines the process of gravity as field potential also correlates to the mechanism of inertia and inertial mass. The point of kinetic energy becoming potential energy due to a transfer between fields provides context. We do not need a separate field to explain them and can easily do it by explaining how they function in a base field of space-time.

***

How gravity functions has been a source of debate for a very long time. The most notable people that have contributed to the subject are Copernicus, Kepler, Galileo, Newton and Einstein. Each have progressed the subject to an even higher level with Einstein's contribution being the last but not one that answers all questions. In fact along with the advancement came more questions and that fact creates gaps which works against his theory as a whole. I believe he came close to the final theory but because he was working it out under the assumption of it being an attraction the direction he took was off course.

I will begin my explanation of gravity at the point where general relativity transitions from rational to irrational. I have to rearrange GR somewhat because it states time dilation is a consequence of gravity. That statement puts the order of events incorrectly for which affects the rational of the whole process. In my view gravity is a consequence of time dilation and that is a consequence of the governing action space-time has on energy to maintain a maximum speed. Einstein was correct to a point but places the order of the process out of step which may be another reason it went off course.

The process of gravity begins with the governing action of space-time to maintain the speed of energy in the universe. The compounded effects of that show up as time dilation for objects of mass, an increase of mass creates an increase in the effect. You could say that the process of gravity begins at the most basic level of space-time.

A characteristic of time dilation for a mass is a diminishing reduction of the effect with distance from it. The word time dilation characterizes a slowing of time so with distance there will be an increase in the rate of time as a function of an inverse square. That means time will increase exponentially as distance increases from the center of the mass. The equation for this phenomenon supports the characteristic that the effect of gravity has no bounds and never ends. Both gravity and time dilation follow this characteristic so the connection should be evident.

Introduction

This never ending chain of relative time of space-time creates a difference of space-time from one point to another such as from one side of an atom to the other. Atoms occupy a volume of space-time and within that volume spans a difference in space-time. The term for that difference is differential space-time and or relative space-time and simply means that there is a difference in the rate of time as well as spatial volume. The word time dilation is misleading in a way since it focuses on time but we should focus on space-time because the two are inversely linked. If there is a difference of time from one point to another there is also a difference of spatial volume which satisfies the link between space-time.

When I state the differences of relative space-time create gravity remember I mentioned the inverse characteristic between space and time. As time gradually increases spatial dimension gradually decreases. Looking at the spatial aspect inside of an atom from the direction of a source of time dilation its volume would be greatest. From that point leading away the spatial volume would diminish as the rate of time increased.

That specific characteristic changes the energy potential in the atoms fields. Think in terms of coulombs law where the closer opposing polarities are together the stronger the field and the further they are the weaker the field. In the case of an atom the largest field would be the electrostatic field between the electrons and the nucleus and is one of a few fields where the imbalance would occur. The spatial imbalance would create a field imbalance and the stronger side of the field has the inherent higher potential. The higher potential propels the atom toward the source of time dilation causing the imbalance.

In a rocket accelerating through space the engines produce an energy potential at one end. Because the energy potential at the other end is lesser it is propelled in that direction. Gravity works in a similar way but obviously with a different mechanism. Gravity is propulsion in the way it functions so it is also obviously not an attraction. When you ponder how gravity reaches out and affects an object keep in mind that an object is simply reacting to the space-time it's occupying. There is no reaching out or any wave or particle spanning the distance to be a factor of gravity. The relative space-time that an object creates and another object reacts to is behind the mechanism of gravity.

A change in time dilation also follows the natural speed limit in the universe. Space-time cannot change faster than the speed of light which is why the effect of gravity cannot occur faster than the speed of light. That is why if the sun were to vanish it would take eight minutes to affect the earth's orbit because it is eight light minutes away. The governing mechanism built in to space-time manifests in multiple ways.

Gravity is simply how an object of mass reacts to a change of characteristic in the fields that create mass by relative time dilation or relative space-time. To give you some perspective on how field potential creates gravity compare in your mind the difference between an object in differential time dilation and also in zero time dilation. In zero time dilation there would be no relative time differences and therefore no spatial differences in an atom. From the perspective of the nucleus to any point of the electron orbit all measurements of time and spatial dimension would be equal. No differences create a balanced field and a balanced field would have no point of higher potential. Compare that to an atom with an imbalance in its fields where within the space-time volume it is occupying there are differences of time and therefore spatial volume. It is that imbalance that with further explanation causes gravitation.

****

The electrostatic field is not the only field associated with an atom. It is the main field between the electrons and the nucleus and minor field between the protons and neutrons. Inside the nucleons the strong force holds the nucleus together regardless of the electrostatic repulsion. That comes from the field between the quarks or the gluon field. The context of that statement is that there are two fundamental forces at work for an atom that act as fields that give it mass. An object existing as a field occupies a spatial volume therefore relative time dilation created by a source will create a potential in their fields. That means since gravity is a mechanism that affects fields of either electromagnetism or of the strong force it is not a separate fundamental force of nature. Putting it simply gravity is a characteristic of potential induced in a field while occupying differential space-time.

Looking at gravity presented here as a characteristic of a field of a fundamental force it shows that it is not a fundamental force in the way functions. The repercussions of this are that we will never find a graviton particle because only actual fundamental forces have their own particle.

Another repercussion of propulsive gravity is that there are only three fundamental forces of nature, the strong force, electromagnetism and the weak force. The simple process for gravity explained here in a way ends up as a unification of forces without any additional effort or design. The context of the unification is more of one that unifies gravity with electromagnetism and the strong force individually. It is not the unification of all three into one. Creating a chapter to cover it would not be beneficial since it is a characteristic of the mechanism of gravity and not a separate factor unto itself. That should speak volumes to the validity of the design of this theory or hypothesis. Explaining gravity the way this theory does has the unification built in and should be self-evident.

Time Dilation

We have Albert Einstein to thank for the introduction of the concept of time dilation. For taking the world to another level in comprehending how the universe functions. The fascination that we have with time I'm sure made his theory about it grasp peoples imagination in the same way as if someone just discovered an alternate parallel universe in our day. Remember there was no radio, television, internet or smart phones in that era and reading was the only way to get current news. The idea that time could pass at different rates in different places is a fascinating concept even to this day. The novelty of that has never worn off.

Einstein later contributed the characteristic of time dilation to his general theory of relativity. The major points of that theory are gravity, mass, time dilation and curved space-time. In my opinion he places these things in an improper order for the conception of gravity by not offering the true genesis of time dilation, the base level of how time dilation manifests. That in a way makes the steps of his process irrational. I believe by rearranging the steps rationally a new outcome for defining gravity presents itself. That new outcome is akin to a higher definition picture which shows us new detail. The old image is still present in a sense but we see it with a different perspective.

***

This theory of gravity has its origins stemming from the simple characteristic of time dilation. In it an object or a mass has the characteristic of a slowing of time due to the regulatory action of space-time. Space-time must create equilibrium throughout the universe by limiting how fast anything can exist or move. It does so by the mechanism inherent in the balance of the relationship between space and time. It regulates something by slowing time for it which is why it's termed time dilation though the word dilation is more of a verb in the phrase.

When time is dilating it characterizes a slowing of time as an object increases in mass. The same holds true when an object with velocity increases in velocity. The title "time dilation" has taken hold in society which is not such a good thing. It does not equate to the whole scope of what is really happening in the process. In order to look at it specifically you have to recognize space and time are one. A more appropriate term would be space-time dilation emphasizing an alteration in those dual factors.

Since time tends to be more of a philosophical idea to many then the same process can be looked at in terms of space with the same confidence. We know time is affected and also know space and time are two sides of the same coin. It is not a stretch to conceive that space changes in characteristic just as time does. Space reacts to mass or velocity by altering its spatial dimension with an increase in order to accommodate it. That reaction shows as a slowing of time or time dilation because of the inverse characteristics of space-time. An expansion of spatial dimension reacts with a slowing of proper time and a contraction of it with an increased rate of time. We normally only recognize the slowing of time because the normal evolution of the universe is the concentration of mass objects.

An object consisting of an unimaginable number of atoms for instance will experience greater time dilation than a single atom. In terms of a spatial aspect space-time is creating a greater spatial volume for it so as to maintain a proper distance for those energies to exist. It is essentially spreading energy out over a larger area. The opposite of that is concentrating energy into a smaller area and if space can be quantized it can only hold a limited amount of energy. You may look at this idea as prevention for a mass to become critical and go supernova. By giving mass more room it prevents a cascade effect from occurring. It should be considered a basic law in the universe.

The regulatory action of space-time will occur for an object increasing in mass as well as an object increasing in velocity. Even though that those are two different phenomenon they have the same genesis of space-time regulation. Because of that we should not really create titles that may be misleading and create confusion due to inadequate placement of the steps of the process. I'm referring to the terms "gravitational time dilation" and "velocity time dilation". At a basic level they are the same thing and gravity is not producing the gravitational time dilation so it is inaccurate. Gravitational time dilation of GR as I mentioned is the characteristic of space-time reacting to mass as a regulator or governor. This may be considered a simple thing but these improper definitions place the order of the genesis of gravity out of sync. One misdirection at the base level leads to going completely off course.

Another way to look at velocity time dilation is if electrons are orbiting inside of an atom around its nucleus at a set speed then they are already regulated to that speed limit. If those atoms start moving through space-time with velocity such as atoms in a spaceship you cannot add that velocity to which they are already regulated to. The speed of the electrons in their orbit will slow by exactly the amount of the added velocity. That slowing is a result of increased time dilation for that atom. They slow to the speed that they were originally set to as a limit which is proper time in that atom.

For reference sake, that atom is one of many in a spaceship on its way to mars. Any energy and structure of energy when at rest will be regulated by space-time to a set speed. Any motion or velocity will induce a characteristic of time dilation to it. Consider the at rest rate of time to be proper time and the velocity time dilation a mechanism to restore proper time to it while in motion. The alteration of time is the governing mechanism to achieve a sense of order in the universe.

***

Theoretical mass is manifested in an object of mass when in motion. It is directly related to velocity time dilation and even though termed theoretical has real effects. It's really called theoretical because those effects come and go from transitioning between an at rest state and in motion state. The correlation is between the rate of time for a mass at rest and the slowing of time of it in motion. The equations and calculations work it out very well but that leaves little for the imagination to ponder.

The gist of the phenomenon centers on the changing of the rate of time. An increase in the mass of planet earth results in the slowing of time. Increasing the earth's velocity in its orbit also results in a slowing rate of time. Since the slowing rate of time is a result and the second example reproduces the effects of the first example the earth is said to have an increase in theoretical mass. It's a kind of balancing act in the equations but even so it does result in what this whole topic is about which is gravity. That is because gravity is directly related to time dilation regardless of how that time dilation occurs or is manifested. What comes with velocity time dilation is an increase in theoretical mass just as with an actual increase in mass it produces a greater gravitational field. A gravitational field is a field of time dilation but more precisely the resulting relative space-time it creates. Gravity is produced regardless of how the time dilation is created. In the following I will toggle between the terms "relative space-time" and "differential space-time". They mean the same but the word differential gives a more specific meaning because it characterizes the differences of space and time from one point to another.

Imagine this example, the earth is now rotating or spinning as well as orbiting the sun. There are two forms of motion creating additional time dilation over that which would be if the earth where completely stationary. At the current state of motion if you stood on a scale and it read 100 kilos that is technically your theoretical weight because it's based on your theoretical mass while in a state of motion. Now the earth instantly stops rotating and orbiting the sun to become completely stationary the reading on the scale would decrease to an at rest value. That is because the additional time dilation of motion is withdrawn as well as the additional theoretical mass added to you. This is an accurate simple example not taking into account the motion within the galaxy or universe or inertia of instantly stopping. This is to emphasize that the characteristic of theoretical mass manifests in real life because it is created by real time dilation regardless of what is causing time to slow.

***

If we are to move forward in our quest to get to the bottom of how things work in the universe we need to be more specific about the terminology so that the correct chain of events can be put into the proper order. It also makes all of the pieces of the puzzle clearer so they can be placed appropriately to show what we know as well as what we don't know. That is the only way we can extrapolate what we don't know into something with greater possibilities. Creating unclear and unspecific terms for characteristics in the universe only creates distance between unknowing and knowing.

***

We cannot tell the difference between a distorted view and real time. We see an irrational picture and try to rationalize it to have it make sense. The problem is these rationalizations that we use don't follow the basics of what we know is fact. That creates an inaccurate picture so that the next picture in the process to try and understand the universe is based on a distorted assumption, not simple truth. That is the place where we are today in the scope of cosmology and if we don't take steps to reverse this trend we will never go anywhere let alone out into our own solar system. That is one of the topics covered in my other book "The Universe 20/20" in the chapter about dark matter among other things.

The whole point of this line of reasoning creates an entirely different direction for general relativity. GR concentrates more on the time dilation as a consequence of a gravitational field of a mass and curved space-time initiating the act of gravitation. I concentrate on the relative time of time dilation which creates changes in spatial dimension. That difference in spatial dimension affects anything that is occupying it and if it exists as a field such as with mass then the field becomes imbalanced and an energy potential is manifested. That potential is the propulsion of gravity.

Time Dilation *"An effect of space-time on space-time to maintain a single speed limit for energy by altering the factors of space and time"*

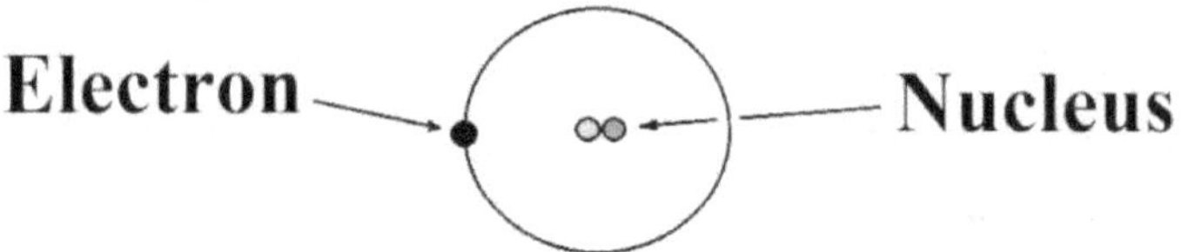

Illustration 1-1; Time Dilation effects in atom formation

Time dilation governs the speed of everything but also allows atoms to form by compounding the time dilation when an electron, proton and neutron are in close proximity. When those are near to each other the time dilation becomes slower than for each individually. Time slows down to allow them to get caught up in each other's field. It should be of no surprise that the hydrogen atom is the first to be created like this and may be the only atom created like this since more complex atoms are created in stars.

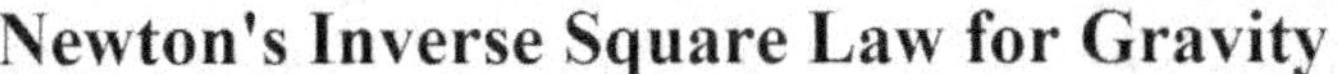

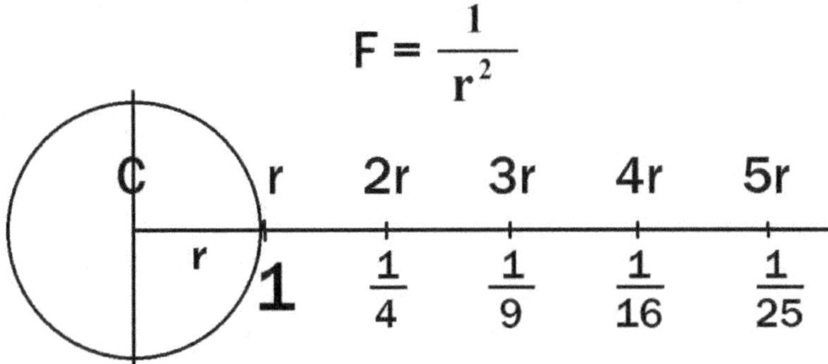

Illustration 1-2; Isaac Newton's inverse square law

This is a representation of how the mechanism of gravity diminishes over distance. This shows the diminishing effect of time dilation and how it creates relative space-time because of that. Diminishing time dilation equates to an increase in the rate of time. The further from the planet the weaker the time dilation and less differential of space-time. Since it is relative space-time that induces an energy potential in an atoms fields the less propulsive gravity is produced with distance from the source.

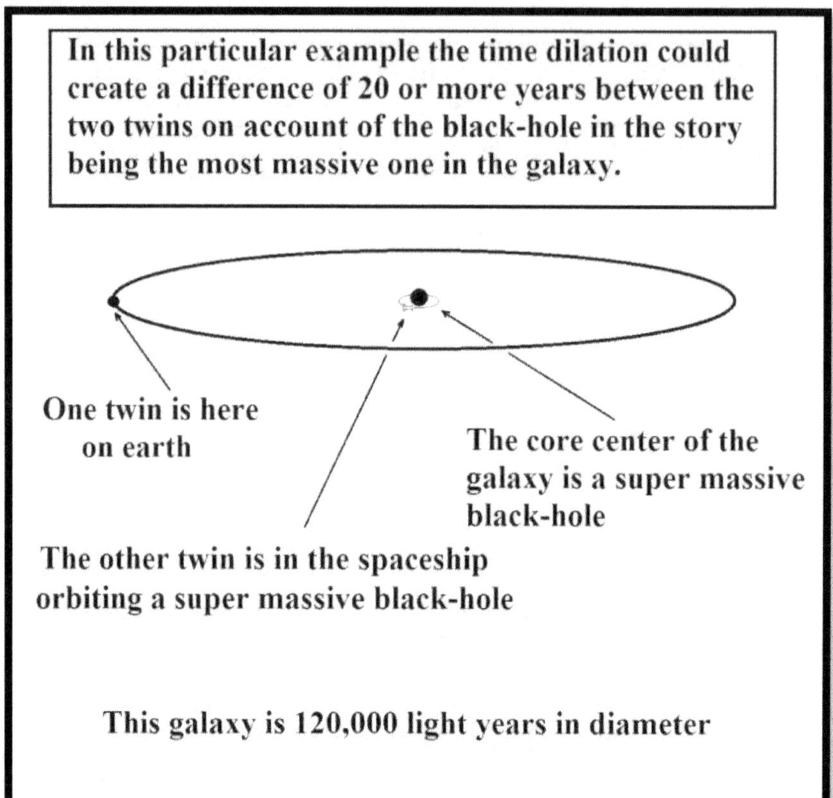

Illustration 1-3: The twin paradox result of relative time

A large mass will produce a significantly large amount of time dilation which is a slow rate of time. The speed of time is different when comparing the center and the perimeter of the galaxy. A pair of twins with one at each location would age independently of one another. One will grow old if at the perimeter location or on earth and the other would age much slower if in an area of high time dilation (slow time) such as near a super massive black-hole. Adding to the time dilation is the velocity of the ship in orbit.

Illustration 1-4; Relative light speed in relative space-time

Time dilation is greatest at the surface of a star, as the light is emitted its relative speed gradually increases. The light after five light seconds has a greater relative speed than after one light second. This is only due to the increase in the rate of time not the actual speed the light is traveling. The speed of light is the same at each point but it is the speed of time creating this relative difference.

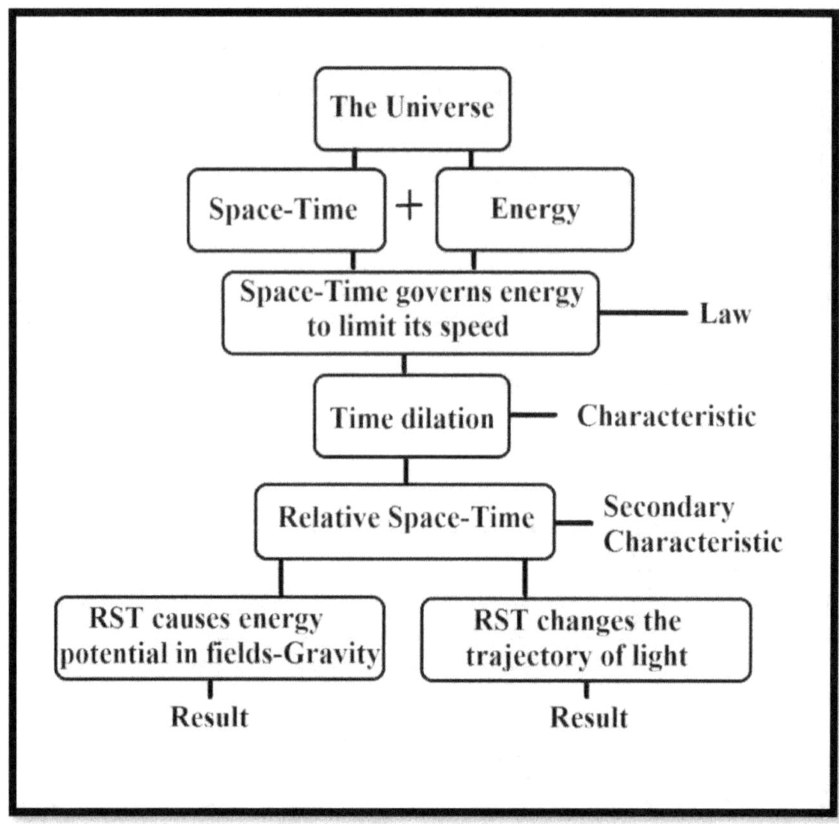

Illustration 1-5; Block diagram of the universe

The purpose of this diagram is to show the active role space-time plays in the universe centering on the law inherent within it to maintain the speed of everything. The law of space-time presents itself as the characteristic of time dilation for a mass and velocity of an object. Two indirect results of that are the creation of gravity and the effect of causing light to bend. Gravity and the bending of light are the result of the relative space-time created by time dilation. The steps are space-time, time dilation, relative space-time then gravity and light bending as separate results.

Relative Time and Spatial Dimension

We have always viewed space and time as two separate factors in the universe until Einstein with the help of Herman Minkowski who combined the two as one and coined the term space-time. Even so if we lose sight of that we think that each exists on its own and can function as individual factors. It's a hard habit to break and sometimes we revert back for convenience. After much pondering of the subject during the writing of my first and second books then writing a few scientific papers in order to summarize the material I've come to the conclusion that we just don't know enough to define each individually. Neither should be defined without considering the other in the context of them being two sides of the same coin, you can't affect one and not the other. In order to create an accurate picture of the universe we must always look at the factors realistically. The reality is that space-time is one factor and not two acting as one. In the yin yang symbol we normally don't see it as two symbols together we automatically perceive it as one.

How Time Dilation Creates Propulsive Gravity

In the previous edition of this book I had written that space is nothing more than a construct of time with time being real and space determined by the rate of time. In that context I was concluding that time is the one and only factor and space a consequence of time and not real. I believe now that was premature and an assumption on my part. I apparently fell into the same trap as many and made a leap or assumption without seeing that it created blank puzzle pieces and separated the rational pieces further. Since then I've had to reset my way of thinking about space and time or space-time. Regardless of which is real or which is a construct or neither it is better to consider their interacting duality when trying to figure out how things function within the space-time fabric. We can acknowledge their inverse relationship and worry about defining them in more detail later.

In the previous chapter on time dilation I spread out among it some details here and there to seed your mind with some basics of space-time. Those ideas should be easily accessible to your subconscious while reading through this chapter so as to ease its comprehension. I'm not saying this topic will be hard to process just that if you have already assimilated it that will lay the foundation for absorbing this chapter. All of the topics in this book and for this theory are very basic. The theory for gravity presented is simple and using known factors with simple foundations. As I've mentioned previously we don't need complex and complicated mechanisms to explain simple things. Gravity is the most basic action second to time dilation in the universe. When a theory is laid out with complexity it's moving further away from reality.

Relative Time and Spatial Dimension

***

When you start focusing and trying to comprehend relative spatial dimension relating to the speed or rate of time you can't help being trapped into a certain way of thinking. In the back of your mind you are always thinking that space and time are separate. At this time we don't know that they are but we do know they act as one entity. Whether one is real or a construct or real and a governor it's pointless to try and go further. The further we separate what we know to be true the less accurate it becomes. Just keep in mind the inverse nature of space-time and if you are affecting one you are affecting the other. That will prevent misunderstandings in the future when learning about the concepts associated with the pertaining topics. Your mind will be programmed with truths and have fewer avenues that lead nowhere.

***

Space and time have an unbreakable connection and there is a little known correlation between them. The correlation is that slower time has more spatial dimension than faster time. When large masses form such as a star time will slow for it by means of time dilation in order to govern it. That time dilation in essence is also giving the mass of the star more spatial dimension at the same time. Because when time is slowed down for the mass it is actually gaining more space, there will be more "space" in between each of the atoms. It may be hard to imagine or wrap the mind around so just consider this characteristic a relative nature of space-time.

It's relative because this is not noticed and we don't see the universe with different characteristics that are a result of this. We see it basically flat, no characteristic that is created by time dilation shows itself as just that. Because the speed of light is a set speed and the speed of time is also set so any changes in the characteristic of space-time don't get emphasized, it blends into a "flat" image of the universe. In some cases that gives us false data when we view something and have no way to explain it because we aren't getting the additional information that's causing the false image. We look out into the universe and see that picture and believe we are looking at the current universe and we aren't. We don't take into account that the universe is sending us the data to make the picture and that data has long since expired or changed. That is only speaking of the time it takes us to receive the data not including differences in the rate of time.

When measuring distances on a very large scale that is done using the speed of light as a unit of measure. The term is called a light year and it's the distance that light will travel in one year. It's a unit of time that describes a unit of distance. It also has a correlation to speed as well as distance because it is regulated by time. If you have a given area in space you can measure it with a beam of light. Since we know the distance light will travel in a given amount of time we can determine the dimension of space. Imagine you are looking at a round sphere in space and a laser is triggered to pass from one side to the other. Let's say it takes one second for the laser beam to make the trip from one side to the other. We already know how fast light travels so we know the sphere is 300,000 kilometers in diameter. Let us look at another sphere next to the first and it's the same size visually. Without any experimentation we assume we already know the size of the sphere.

When the laser shoots a beam into that sphere it takes five seconds to pass through it. Wait a minute it's the same size as the first how could it take five seconds? The answer to that is there is a black-hole nearby causing only the second sphere to experience extreme time dilation. The time dilation is slowing the rate of time in the sphere limiting the speed of the laser as it passes through it. They both looked the same from our perspective yet the second sphere the laser took five seconds to pass through. That was five times longer than the first which in turn makes it five times larger. Since the speed of time is a constant and the speed of light is a constant and we use light to measure distance in the sphere it makes the sphere at least five times the spatial dimension as the first sphere. The characteristics like that are what make it hard to see the real picture of the universe. We cannot always trust what we perceive things to be and that example tells us slower time has a larger spatial dimension.

***

Time will slow for energy or a mass (a structure of energy as a field) by it being a regulatory action in the fabric of space-time. Time is slowing and or space is expanding to govern them not the other way around with them causing time to slow. You could say a mass will slow time but that is an effect not a cause and it's important to reveal the process correctly and in order to create a strong foundation. Since space-time is part of the universes foundation and its mass was created in part by space-time in my mind I say it the way I see it. It's a small detail but has a bearing on creating an accurate view of the process. Knowing that space-time has a governing action at the most basic level is one step of this process and may help us understand some other aspect of the universe further down the path. If we keep things accurate at every stage of the process we can stay on the right track.

***

It may help you to think of space and time as a window screen, the grid you see is time and how close the lines of the grid are together represents space and both together represent space-time. The metal grid is real and tangible and the spaces in between you wouldn't see if it weren't for the grid of the screen. If you expand the grid you are speeding up time and slowing it when compressing the grid. It's easier to image if you think of each square to be one light year in distance. When time is slower you can fit more grids into the same space as the faster time giving it relatively more spatial dimension. If you speed it up by expanding it you reduce the amount of spatial grids in an area.

The maximum speed of time is the speed of light. If you speed up time to the speed of light you lose spatial dimension because you lose the lines of time that represent space. That is the fastest speed of time possible in this universe, not only does space-time with time dilation hold everything to the speed of light, space-time is also held to that maximum speed because of spatial limitations. That notion would be the idea when trying to conceive of a quantum of space-time and working out the details of its fundamental nature.

On the other hand if you compress the lines of time to show how slow it can be represented then at some point the lines become too close together and you lose what represents space. Because of this you have to imagine that you can compress the lines of the screen together only so far and it still resembles a screen and any further it doesn't. That analogy represents that time can slow down to a certain point but not to zero. In this example a zero rate of time equates to infinite spatial dimension and a rate of time equaling the speed of light represents a zero spatial dimension. If there was a zero speed of time the universe would actually be infinite but theoretically static which is why it cannot be because we know it's not static.

Those are only relative characteristics for the energy that exists in the universe in its frame of reference. It can never be said that anything of mass can be in a zero rate of time since that would be like a missing tooth on a gear inside of a clock, the clock would stop and would never run again. If the speed of time reached the speed of light there would be no space between anything, the universe would have no space. That characteristic is most likely to only happen when one universe transitions into another universe described in "The Universe 20/20 a new and unique hindsight view" when a single remaining black-hole absorbs its host universe.

The preceding was a precursor on the topic and now I'll go into how time and spatial dimension are steps in the mechanism for gravity. At this point the steps consist of time dilation, relative time and relative spatial dimension. The relative time and spatial dimension are the result of the dissipation of time dilation through space-time "relative and differential space-time". Time dilation of the earth does not stop at the boundary of where mass ends. Space-time characteristics lead to a slow drop off of the effects. The time dilation of the earth dissipates following an inverse square function, time increases with distance exponentially. The characteristic of this effect never ends which parallels the characteristic of gravity. Time dilation associated with a mass object creates a never ending span of relative space-time.

To be specific about how spatial dimension is part of the process for gravity it is about the characteristic that time dilation imparts to space-time. The dissipation of time dilation creates relative time extending away from a mass. The rate of time for the earth at sea level is slower than at one mile above sea level. In between those two points are a multitude of points where the rate of time is different than both. In between all of those points is another multitude of varying points. There are no distinct points but a variable change of relative time. I suppose that if space-time can be quantized then each quantum of space would have a distinct time and spatial value.

Here is something for your imagination to ponder that may just focus the topic into a concentrated point of awareness for you. I will refer back to the governing action of space-time that creates time dilation to set this up. Imagine a quantum of space-time that can be considered its smallest indivisible unit. We know that time dilation occurs within space-time when energy occupies that space-time. I believe that space-time is naturally reacting to that energy by expanding its spatial dimension. It's a natural process so that space-time can never normally reach a point of critical mass which would initiate a premature black-hole event.

Space-time will always continue to react to the addition of energy and increase in spatial dimension to accommodate the increase of energy. That prevents the quantum of space-time from ever filling up to the brim with energy for lack of better words. If it did fill up to the brim that would trigger a natural mechanism built into space-time and create a black-hole. That would leave the formation of black-holes to be only by means of a collapsing star where the energy as mass is compressed to a point where a quantum of space-time becomes over-saturated where the direction of time inverts in the pocket of space-time as its birth. I do not use the term singularity here because the definition created for it makes it more fiction than fact. A black-hole is more appropriate and exact with the characteristic of it being no longer in the same timeline of the universe but a small part of the next incarnation of one. The only difference between the space-time of the universe and a black-hole is the direction of time.

Whenever an area of space-time has an increase of energy it increases in spatial volume. At the same time it is increasing in volume the rate of time for it is decreasing. In a roundabout way that correlation is proven by the acceptance of time dilation but the mechanism has not been fully clarified for how it affects space but time only. The process of a quantum of space-time reacting to energy and increasing in spatial dimension is how time dilation manifests as a governor. The flip side of the coin is as that takes place the rate of time decreases hence the term "time dilation". The term is focused on time which is why the effect on space doesn't normally become apparent because it isn't looked for or is assumed not to have any.

If you research the topic of the expanding universe you will find that the main reason for it is specifically related to an expansion of space and not the velocity of moving galaxies. That would set the precedent for the effect and give credence to it at a much smaller scale. That would be at the level of atoms, nucleus of atoms and nucleons when they occupy the relative space-time created by a mass in the context of gravitation. If it is commonplace to accept spatial expansion on the grand scale why not at the smallest scale?

It may be occurring at a small scale and simply not recognized or passed off as a different phenomenon. For instance if we fill a pot to the brim with water and apply heat as energy the contents will expand and overflow. This represents the water as energy in the form of atoms and they fill the pots spatial volume completely to the top. The energy value of the atoms expands the spatial volume to a certain value at room temperature. It is a commonplace fact that atoms are primarily spatial volume and the energy value is constituted by the particles. The pot is full at room temperature and by applying energy as heat to it that is increasing the energy value in the combined quanta of space-time of its volume. Each quantum of space-time reacts to the increase in energy and expands the spatial dimension of the contents of the pot for which the limit of the pot can no longer contain because the volume of the pot stays the same.

We see a precedence set for the whole universe and a simple explanation for the small scale but the inverse relationship between space and time has not specifically been defined in this way to explain how gravity functions. Now it has and it creates the foundation for how this mechanism of propulsive gravity manifests. I believe it will be these specific points of time dilation and relative space-time that will prevent this hypothesis from being accepted even though they aren't that much of a stretch from the norm.

***

We know that time is relative with distance from the earth now all you have to do is factor in the inverse relationship between space-time. With an increase in the rate of time comes a decrease in spatial dimension. That varying increase in the rate of time creates a varying decrease in spatial dimension. That is essentially a gravitational field because it sets up the mechanism of gravity to manifest in objects of mass that are occupying it. It's pretty much like that with no need to over define that stage of the process, it's that simple.

At this point the mechanism for gravity begins at a fundamental level in the universe which is the governing action of space-time and creates time dilation. A characteristic of space-time then creates a field of relative space-time which is defined as differences in the rate of time and spatial dimension. It may be referred to as "differential space-time" or "relative space-time" to simplify the characteristics in the mind. Those characteristics then apply to any object of mass occupying it. When an object of mass exists within a field of differential space-time it changes the inherent balance that normally occurs within its fields. That changing of the balance equates to gravity manifesting within the fields of the object of mass. An imbalance would create a lower and higher energy potential.

Gravity is more of a local frame of reference phenomenon. Meaning it manifests by reason of an object of mass's fields due to the differential space-time that affects it or it is occupying. In comparison to being affected at a distance by some sort of wave or particle that travels at the speed of light. Time dilation affects space-time and space-time cannot change faster than the speed of light which is why gravity cannot manifest faster than the speed of light.

Relative space-time *"The property of two areas of space-time that have a differential or difference in characteristics between them"*

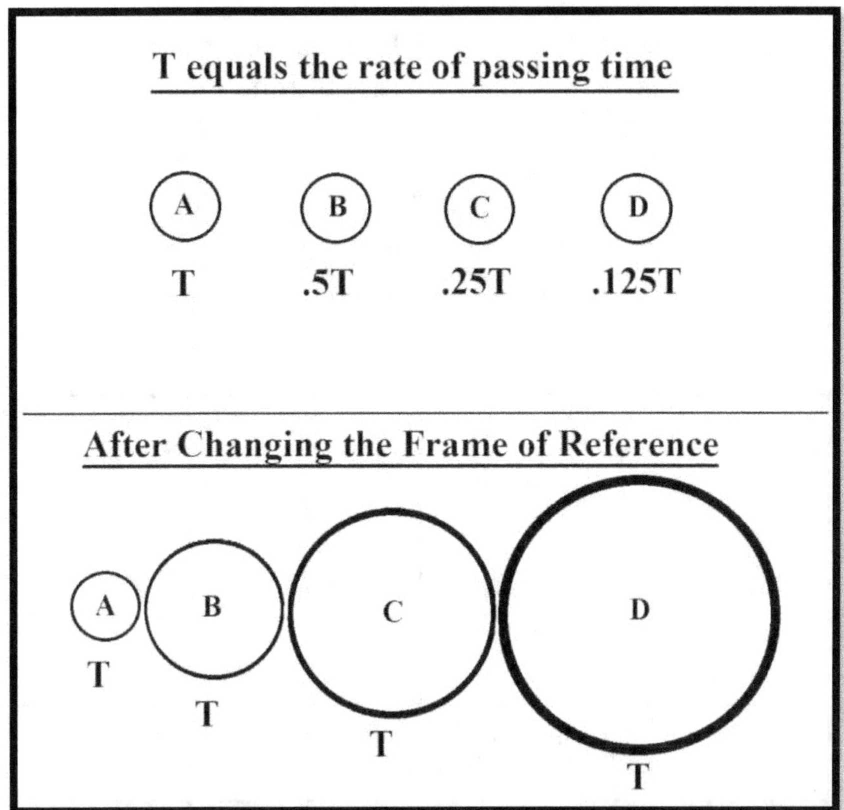

Illustration 2-1; Relative time verses spatial dimension

This is a simple example of spatial dimension relative to the rate of time. When the rate of time changes so does the volume of spatial dimension. The inverse relationship of this effect is why space-time is considered two sides of the same coin. Time dilation creates relative space-time as it dissipates from a source and that creates differences of spatial dimension. Even though it's called time dilation the effect on spatial volume is not typically imagined because the terminology is one sided.

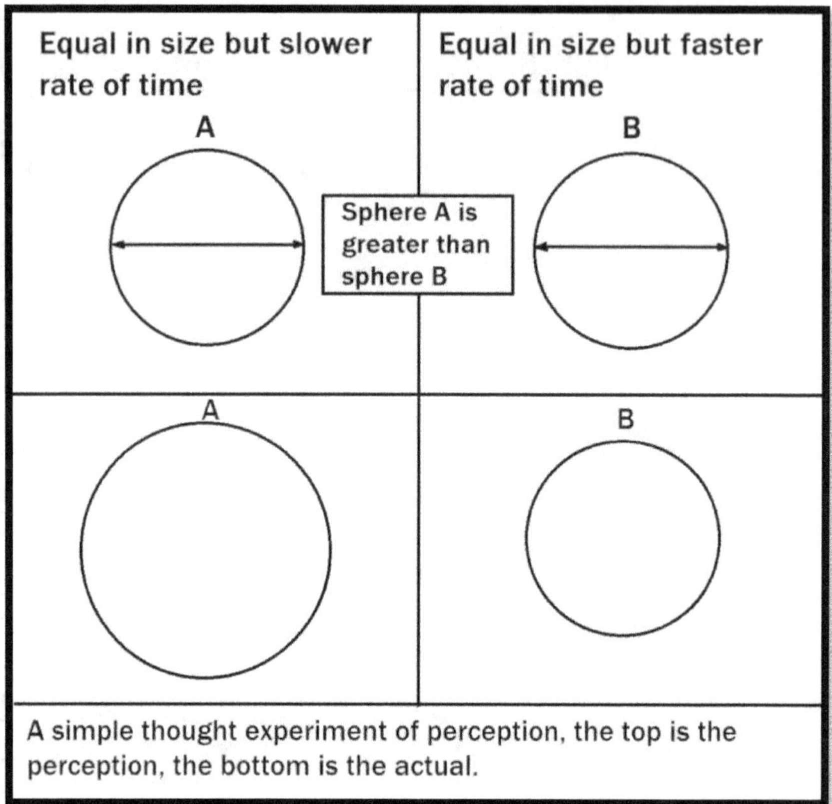

Illustration 2-2; Relative space-time perception

Spatial dimension is relative to the rate of time. When the rate of time is slower the perceived space is smaller and the actual space is larger. When the rate of time is faster the perceived space is larger and the actual space is smaller. We know that black-holes exert an extreme effect of time dilation on space-time. The spatial dimension at the event horizon is the largest and it decreases exponentially as you move away from it. Then spatial volume would be greatest inside it and the outside not representative of that fact.

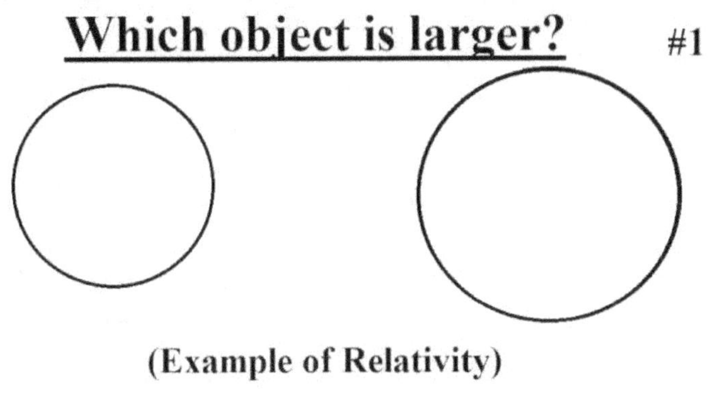

Most people would look at them and say that the right sphere is larger because that is the only perspective available to them. But since time has a corrolation to spatial volume that would be an assumption. Answers derived from multiple perspectives including time have a stronger foundation.

Illustration 2-3; Relative nature of space-time part 1

This is a simple thought exercise of space-time and perspective. It is designed to ask more questions in order to determine what is true compared to what is perceived. With Einstein's special theory of relativity space and time are relative and can be very different at one location relative to another as well as relative to each other. If you use relativity as a guide then you will be less likely to end up making an assumption.

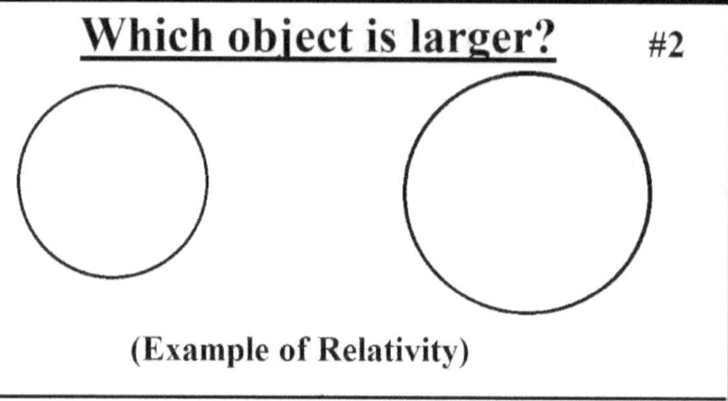

These two objects just happen to be galaxies and the one on the right the rate of time is twice the speed of that on the left. Which means it takes light longer to pass through the left one thus showing it has more spatial volume. We cannot trust what we visually see of the universe and must rely on data from multiple perspectives.

Illustration 2-4; Relative nature of space-time part 2

More information is given here in order to comprehend what is real. When observing something very large such as a galaxy the differences in time are much more pronounced than in our local environment. We take that for granted and assume those characteristics are insignificant everywhere else which leads to misunderstandings about what is factual and what is perceived. It leads to receiving false data and then to assumptions and finally to false theories. The sphere on the left is actually larger due to a slower rate of time.

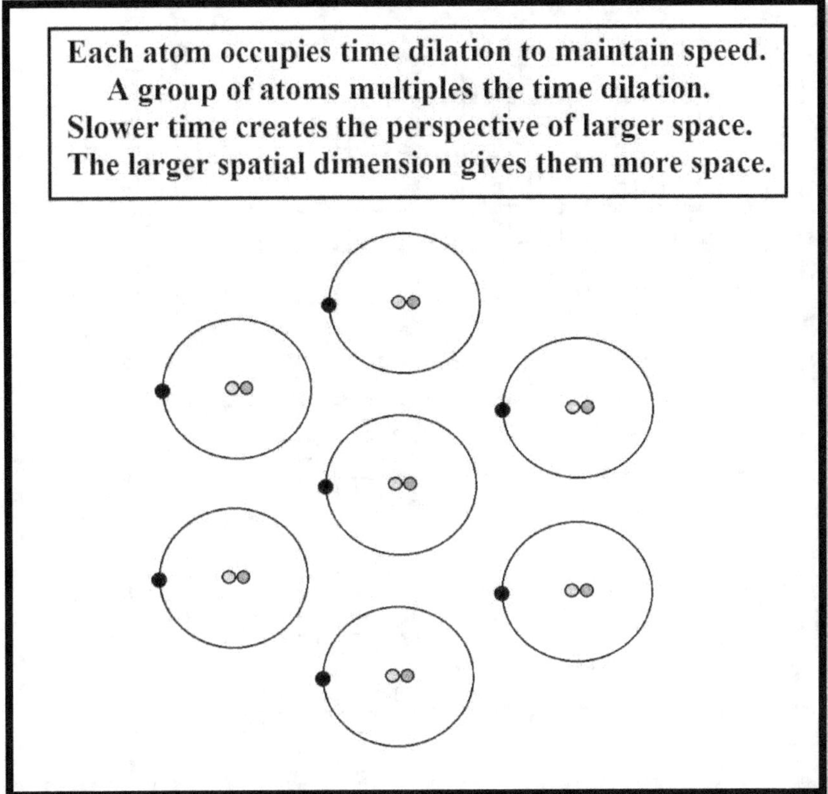

Illustration 2-5: Slowing time increases spatial dimension

Space-time governs mass to a specific speed, this is called time dilation and time will slow down for mass in the area it is occupying. The larger amount of mass in an area will compound the amount of time dilation and time will slow further. The slower the rate of time the greater the spatial area thus giving the mass more space to keep the power of the atom in check or simply making more room to prevent an over accumulation of energy within a given portion of space-time.

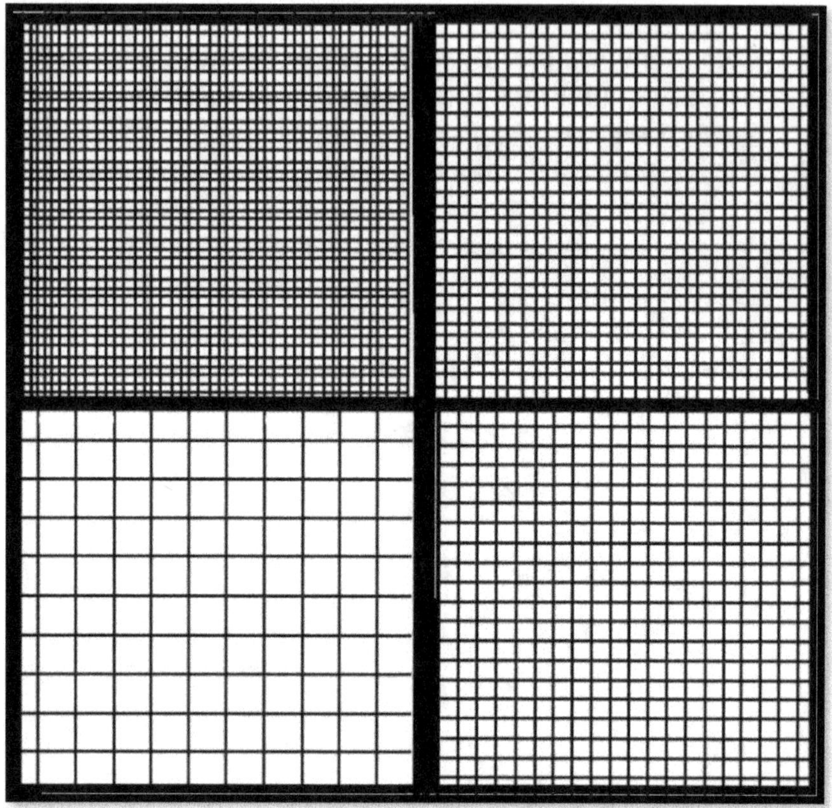

<u>Illustration 2-6; Analogy of time verses spatial dimension</u>

This shows an analogy of time verses space, the closer together the lines the slower the rate of time. From top left CW each section increases the rate of time by 2. Each square unit in any of the sections is one light year. Time is only expressed by the lines but you still perceive what is seen as space. When time is slower there is more space. You visually see the same amount of space in each section but by virtue of the rate of time there is many more times the space in the upper left than the bottom left.

Propulsive Gravity from Energy Potential

Gravity has been falsely identified as being an attraction throughout history. As a result we have never had an adequate understanding of the process that leads to gravitation. The assumption that it's an attraction leads to theories that start off on the wrong foot and therefore create weak foundations. The description given in this book came together when characterizing gravity as energy potential and that potential as inducing propulsion. The process is detailed with Occam's razor as a frame and what lies inside the frame is a basic and simple rationalization between each step of the process. The end result and the sum of its parts create a panoramic view which shows gravity to be a characteristic rather than a fundamental force. Though not fundamental it is still a force, it is just induced by a characteristic.

***

How Time Dilation Creates Propulsive Gravity

Gravity is the result of field energy potential created between the characteristic of relative space-time and the fields that an object of mass consists. Those fields pertain to the fundamental forces of electromagnetism and the strong force. The strong force working inside of the nucleons and the electromagnetic force working between the electrons and nucleus as well as between the nucleons though over powered by the strong force in that context but still there. It is the fields of an object of mass that have the power to motivate them through space-time. How that occurs is simply having an energy potential develop within the field which then propels it. The energy potential is directly related to the strength of the relative space-time which applies the appropriate imbalance to the fields. The imbalance is the potential energy being manifested.

In the context of the process or mechanism gravity is not a separate fundamental force of nature. It is instead a characteristic induced in an object of mass, more specifically in the fields that represents its mass. The mass of an object is technically just the field it is made of. Since gravity is the propulsion of a mass by reason of an imbalance in the masses fields it would be considered a characteristic rather than a fundamental force. That aspect was not predetermined at the genesis of this theory the theory had levels of progression that lead up to that conclusion. It was not foreseen until the theory leveled up and the dust settled.

It makes a big difference in my mind about this theory when comparing it to where I was when I thought the previous version was as far as I could go. Not only does this line of reasoning explain the mechanism of gravity but goes as far as to unify it with other fundamental forces. That broad scope alone should allow this to be looked at seriously and not ignored as another crackpot idea.

***

The heart of the theory in the first edition was based on relative electron speed. An electron will orbit its nucleus and pass through the relative time the atom was occupying. That would create a higher potential in the electron field or electromagnetic field. The atom existing in relative time would have a slow rate of time on one side facing the source of time dilation. On the opposite side the relative time would be greater creating an imbalance of time leading to an imbalance in the field. The relative faster moving electrons would manifest a higher potential in that portion of the field and lead to propulsive gravity.

There was a stepping stone built into that theory that I was not aware of at the time and was there because it was an avenue I didn't fine tune. In a sense I did recognize it but didn't have the tools or conceive of them to take it further. It was the relationship between time dilation and varying spatial dimension it produced. In a way I did what Isaac Newton did and leave it up to the reader to conceive of what effect it may have as he did about how gravity functions.

When I published this book the first time I had thought I had gone as far as I could go with the theory. In the following months I was getting feedback to the effect of "is this peer reviewed?" and the answer was no it wasn't. Because of that I looked at that as something I needed to accomplish and began writing multiple papers. Even though I had no success getting a journal to publish I did look at that process as a turning point. Going to the next level and progressing the scope of the theory further to where it is today. Though I wouldn't say it was one single act, it did take more than one writing of papers to proof read and edit that created a progression of the iterations. The final iteration is what you are reading right now and in my mind I cannot see where it could go any further. I feel it is at a plateau where it will remain until we learn more about the true nature of fundamental space-time and its scope will widen further.

The point at which this was taken to another level is when I started looking at how the differences of spatial dimension would affect the mechanism of the process. When I started to incorporate the fields of an atom into my thought process the pieces started to fall into place. The differences of spatial dimension clearly created differences between the electrons and nucleus on opposite sides of an atom while being affected by differential space-time, that I was aware of but left it at that. When I started focusing on that characteristic and seeing it as a field I made the realization that it created a difference of field potential within the atom as a whole. It ties this theory of gravity to a previously known and well documented aspect of field physics.

In the context of that knowledge a field where its opposing particles are closer the field would be stronger. A field where its opposing particles are further the field would be weaker. In an atom you would look at the field as a whole and the differing distance created by relative space-time imparting an imbalance in the field itself. In the slower time the distance is greater than that of the faster time. In the faster time the distance is lesser so the opposing particles are closer. In the atoms field as a whole the closer the particles are compared to the opposite side where they are further apart an energy potential would manifest. The field in faster time is stronger and has a higher energy potential than that of the side facing the source of time dilation. The field of physics I referred to that sets the precedence for this characteristic is pioneered by Charles-Augustin de Coulomb. This is another example where all of the steps of this process have a previous known precedent. When I say there are no leaps of faith within this mechanism for gravity that means the pieces of the puzzle fit without gaps and with a rational progression.

***

The main examples given for the understanding of the process have been primarily looking at whole atoms. That may be the largest volume field but not the only one. The nucleus has both the electrostatic and strong force at work within it as the mortar between its bricks. The electrostatic field is at work between the nucleons and the strong force at work inside of the nucleons. These two forces are working against each other but with the strong force being more powerful takes precedence. The nucleons hold together because of that power despite the electrostatic field repulsion. Regardless of how the fields function they are fields and the mechanism of gravity creates potential in them if they occupy spatial dimension within relative space-time. Since I am no math wizard I have not tried to figure out which field creates the most energy potential. It could be the electron-nucleus field of the atom as a whole or the electrostatic field of the nucleus or the strong force field in the nucleons. It could be one or all of these fields due to the fact that gravity is such a weak byproduct. It may be weak due to the fact that it is only a characteristic of a fundamental force and not one itself.

***

At this current time we apparently do not have an adequate theory of gravity that all scientists can agree upon. If we did there would be a general consensus and uniform agreement across the board with no outlying theories of gravity being offered but that is not the case. The main theory of gravity that is considered as the most accurate does not explain everything out there in the universe. People come up with all kinds of different, complex and far out there theories to try and explain the broad range of varying discrepancies that plague cosmology. I believe that is the opposite of how to come up with answers. I believe that following a simplistic path with common denominators to link knowledge to facts is the way to proceed.

There is no need to create a modified theory of gravity "MOND" to explain away certain characteristics in the universe that don't fit into a standard theory. The reason a modified theory of gravity would be spawned is due to the factors that make it necessary to create one have no basis in fact. I'm saying that certain main cosmological theories that exist today only exist because they haven't been disproven but also haven't technically been proven. They simply exist as placeholders to fill in the wide gaps until something more tangible comes along to span all of the missing pieces and connect other pieces together. Even so when a main theory of the universe gets accepted by society it tends to stay excepted even if it is wrong because it's impossible to dispute.

The main theories I'm referring to are the big bang theory and the theory of dark matter and dark energy. I also don't agree with the conclusion of general relativity because I believe it is too abstract. It starts off well but it goes off on a tangent when it starts becoming abstract. I believe there is too much emphasis on the equivalence principles used to connect gravity to other mechanisms with a direct correlation being established. The problem with that is analogies don't always tell the whole story so you cannot equate them to the same mechanism.

With the big bang theory there is a big "if", if the big bang theory were correct there would be no need for dark energy, so one false theory creates another. I believe the big bang theory, dark matter and dark energy are false. All three of those falsities make it necessary to fudge a theory of gravity into a MOND, divergent or any theory that needs a complex explanation for gravity. My first book on the subject explained in detail the real mechanisms behind how the universe functions. It should be of no surprise that space-time is the main governor of the universe and any explanation of it will be based on it as a main constituent.

***

The origins of this theory of gravity came about while I was trying to rationalize how an atom would create an attraction to another atom through the medium of space. I looked at the process in a step by step way and I could not get the process to function without making giant leaps that left a lot of specifics unknown. A quality theory of gravity or anything should have nothing left unknown or left to the imagination to fill in any gaps. I continued to work on this problem in my thoughts and mind and eventually gave up on the attraction via space connection. I could not conceive of any possible way that an atom could grab, pull and attract each other or pull on space. That forced me to look at other ways the process of gravity could be manifesting itself between two objects.

I thought to myself what else exists between every atom and object of mass, what connects them that can transmit or produce the process of gravity. I started to focus on time dilation as a means to connect one mass to another, I pondered how that process would function. Even though it was a big moment for me I only have a vague recollection of conceiving the process of how gravity works in that respect. I remember it felt like a serendipity moment and was excited to have finally figured it out but I still had to put it through the wringer in my mind to make sure it was viable. That was the origin of the version that was dependent on the relative speed of electrons in atoms which I thought was the final conclusion.

I figured out that time dilation between two atoms which is diminishing exponentially causes a potential force to be created in an atom by the opposing atoms time dilation. Since an atom is occupying the others field of time dilation and it is diminishing in strength the atom is in a differential space-time field. When an atom occupies an area of space-time that transitions from slow to an increasingly faster rate the electrons speed will vary in their orbit as they pass through the slower and faster time. The electrons will have a relative slower speed in their orbit on the side of the atom where it is closest to the gravitational field (the strongest time dilation). Also they will have a relative faster speed on the side of the atom that is opposite because it has a faster rate of time and facing away from the source of time dilation.

This causes a difference in relative speed of the orbiting electrons thereby creating the potential difference in energy or force. That force is directed towards what is creating the time dilation that is inducing it. Even though the electron is always orbiting at the same speed technically, by reason of perspective and the atom being a unit as a whole it reacts to the potential energy caused by the difference in the speed of time. This line of reasoning even though technically a possibility for gravity was a stepping stone to go further. It had put me on the right track in understanding that gravity is propulsion and manifested because of energy potential. At that point I was just not fully aware of how much further I could progress that.

To reiterate in the original process I proposed the electron will always be moving at a fixed speed while orbiting through the slower and faster time of time dilation relative to itself. Relativity plays a part here because even though the electron has a set speed whether in slow or fast time it is the relative speed of it that causes a potential energy within the atom. In the slower time the electron is moving near the speed of light just like it is in the faster time. In the faster time it is moving faster relatively so that gives the electron a higher relative speed than in the slower time. That higher relative speed is perceived by the atom as a whole as a greater potential energy on the side of faster time which induces a force on that side. In the field of the atom any differences in the field affect the whole atom which creates a potential energy within it. The differences of time created by time dilation affect the atom in a real way even though they are only relative differences.

***

The general consensus between both aspects of the first and second editions of the theory result with the same outcome. The characteristic of gravity is the same as what happens to an atom as anything else you apply a force to the side of, it starts moving. Since the force is applied to the side in faster time then the atom or object will move towards the source creating the time dilation or towards the direction of the slower time dilation.

The potential energy that is induced is a propulsive force and not an attraction mechanism so an atom will be propelled towards another atom and vice versa. The atom does not sense the other atom it only reacts to the field of relative space-time it is occupying created by time dilation. You could say it is reacting to the other atom by reason of its time dilation and since the atom has no senses it is just being propelled through the universe by the currents of relative space-time. "Gravity is just a matter of potential energy which tells an atom which direction to move"

***

In space-time when there is no gravitational field there is no time dilation and therefore no differential in space-time. With no time dilation then the atom has equal rates of time and spatial dimension at all points from the center outward. No time dilation means that the speed of time is the same throughout the inside structure of the atom and that means the spatial distances are also equal. That places the nucleus in the direct center of the atom. The electrons orbit at a uniform rate around the nucleus. That creates a relative and perspective neutral potential energy in the fields of the atom as a whole.

When there is an area of a gravitational field which is essentially varying time dilation (relative space-time) it changes the characteristic inside of an atom. Since it changes spatial dimension by more space in slower time and less space in faster time then the nucleus is no longer centered and is offset towards the faster time. The correlation between nucleus placement in zero gravity or a gravitational field shows that the more offset it is the stronger the force of gravity and vice versa. That is also correlated to the strength of the differential between the areas of the field (the difference or amount of imbalance). That makes it easier to comprehend in your mind how this mechanism works.

***

It was a big revelation to me that the force created in an atom that is considered gravity is a propulsive force. It is easy to see why all of this time and all of history we saw it as an attraction process. When we see an apple fall from a tree we only see the apple in motion from the tree to the ground. From that perspective of being on the earth we see the apple falling toward us as though the earth is attracting it downward. The earth doesn't move so we assume the earth is attracting the apple.

If we were to climb the tree and look down at the apples and see one fall then it is moving away from us therefore changing the perspective of the thought process. If we had gotten another perspective on the phenomenon of gravity our understanding may have taken another path. If Newton had simply climbed his apple tree to get a second perspective of the same phenomenon then he wouldn't have concluded that gravity was strictly an attraction.

If we were to go into outer space and place two apples at a distance of a foot apart and observe what happens we would see the apples move toward each other equally. With that information we could say that the apples attracted each other. But also given that information we could deduce that the apples simply moved toward each other and were propelled to do so in some way. The act of conducting this experiment would leave us with two possible causes for gravitation rather than an assumed one cause. If we had those two choices to pick from throughout history then when we ran out of possible ways to rationalize gravity as an attraction then we would have moved on to working it out as a propulsion. It's a shame we got side tracked for so long because we never got past the initial assumption.

***

It goes without saying that the end result of gravity is force. Force is why a meteor creates a crater when it impacts the moon. Force is directly related to the objects mass and how fast it's moving because it's related to velocity and acceleration. Momentum is also related to force if you interchange velocity for acceleration. When a meteor is falling gravity is applying a constant force to it which is why it's accelerating every second. The meteor is adding to its speed every second it's falling so gravity is a constant force being applied. Even when an object is on the ground a constant force is being applied, that force is always "on". That is why an object rests on the ground because the force of gravity is pushing it toward the ground.

***

It has been somewhat of a challenge to interlace the new updated and revised version of gravity into these pages. I didn't want to start from scratch because the change is really only the end result. But with that said I did want the book to represent the additional awareness and experience I gained after the fact. I went through the book word for word as I proof read and edited it to be sure it parallels my most current thoughts. All of the foundation remains the same with time dilation and relative space-time and all I have to do is not actually change the theory but to change the focus. Change it away from gravity being a result of relative electron speed even though that may have something to do with it to what relative space-times effect on an atom does to its characteristics. To condense all of the characteristics of the revised process into a single list here is the newest version of the mechanism that produces propulsive gravity.

1) Space-time governs energy and therefore mass to a speed limit of the speed of light, time dilation is the result because that is the mechanism.

2) The effect of time dilation gradually decreases and never reaches zero. A decrease of time dilation equates to an increase in the rate of time with distance.

3) An object of mass will occupy the dilated space-time and exist within an increasing rate of time or relative space-time.

4) Using the center of an atom as a reference point each half of the atom will be occupying differing rates of time while in a gravitational field of relative space-time.

5) One half will have a slower rate of time and the other half a faster rate of time.

6) There is an inverse relationship between space and time as a factor to their existence as two sides of the same coin.

7) When time is slowed spatial volume is increased. When time is sped up spatial area is decreased.

8) That relationship creates a difference in the distance between the individual particles of an atom within time dilated and relative space-time or a gravitational field.

9) The difference of distance creates a difference of energy potential in the fields that constitute an atom and nucleus.

10) The energy differential creates a higher energy potential in the area where the field is stronger which is the faster rate of time spawned from the lesser spatial dimension.

11) If you look at an atom as a scale with the nucleus the fulcrum you will have the same charge between the electrons and nucleus in flat space-time. In relative space-time the distance imbalance creates a field imbalance. The further the charge the weaker and the closer the stronger. The individual charges or group of them are the same but the difference of distance is the key. The scale will tip toward the higher potential metaphorically.

11) This phenomenon is known as Coulombs law and pertains to Coulombs equation to calculate the energy potential. It is created by the difference of field potential in each half of an atom as seen facing and opposing the source of time dilation.

12) The inducing of the energy potential within the atoms fields creates a force of propulsion that is better known as and incorrectly termed gravitational attraction.

13) Gravity is not a separate fundamental force of nature it is a characteristic due to its initiation as a result of an imbalance within fields that create mass.

That is the condensed version of the simple process. You must conceive of each step before moving on to another for the process to make sense. If you can conceive of each step and look at the process as a whole you can look at gravity in a whole new light. The new process you conceive of as gravity takes the topic to another level with an assimilation of gravity having to do with fundamental forces but not being one on its own. That changes gravity from being a fundamental force to being a characteristic of one. This opens the door to going above and beyond where we are now in our understanding of how things function in the universe. That includes creating new technology for the purpose to simulate the force of gravity. In doing so we can take our space craft to a new level up and into space because they can be propelled by inducing in them the same characteristic that is created naturally by time dilation to hold them down.

Propulsive gravity *"An imbalance created in the fields of an atom that constitutes its mass that results in developing a difference of potential"*

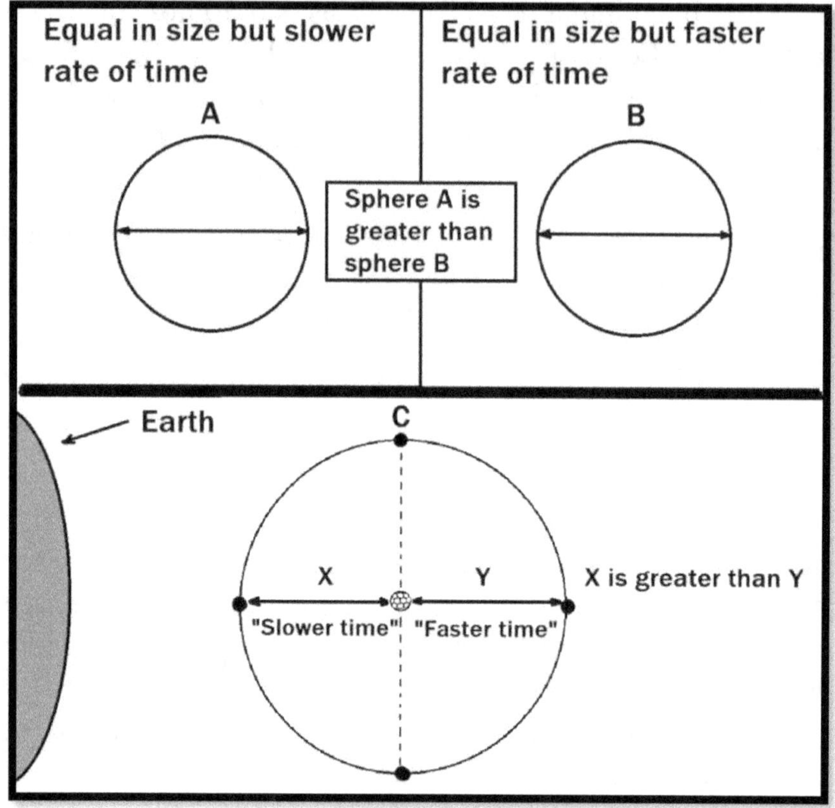

Illustration 3-1; Relative space-time volume in atom

The top half of this drawing shows a correlation to relative space-time and volume in relation to the rate of time. The bottom half shows how that effect plays out inside of an atom while occupying relative space-time. The electron will have a constant speed relative to itself as it orbits. But in the faster verses the slower rate of time it will have a higher relative velocity. The higher relative velocity equates to a higher potential energy causing the atom to be propelled toward the object creating the relative space-time. A cause for gravity?

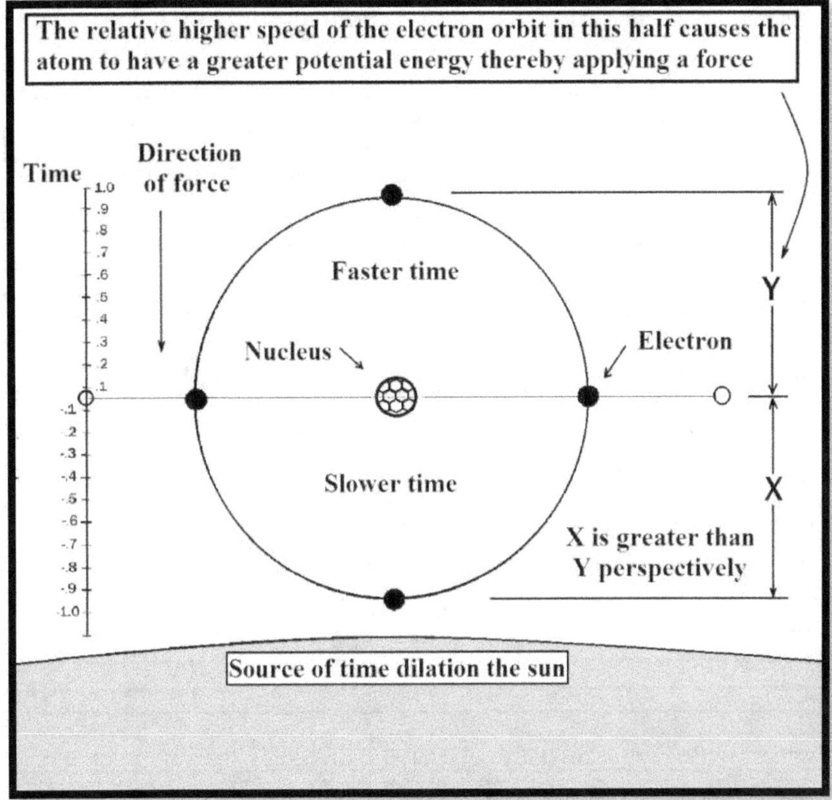

<u>Illustration 3-2; Relative electron speed as energy potential</u>

This illustration emphasizes more the characteristic of the relative speed difference in the electron orbit. The slower time closest to the source causes the electron to have a slower relative speed verses the speed of it in the faster half. The relative difference equates to a potential energy difference which equates to a force being applied to the side of the atom in faster time which propels it towards the source of time dilation. A cause for gravity?

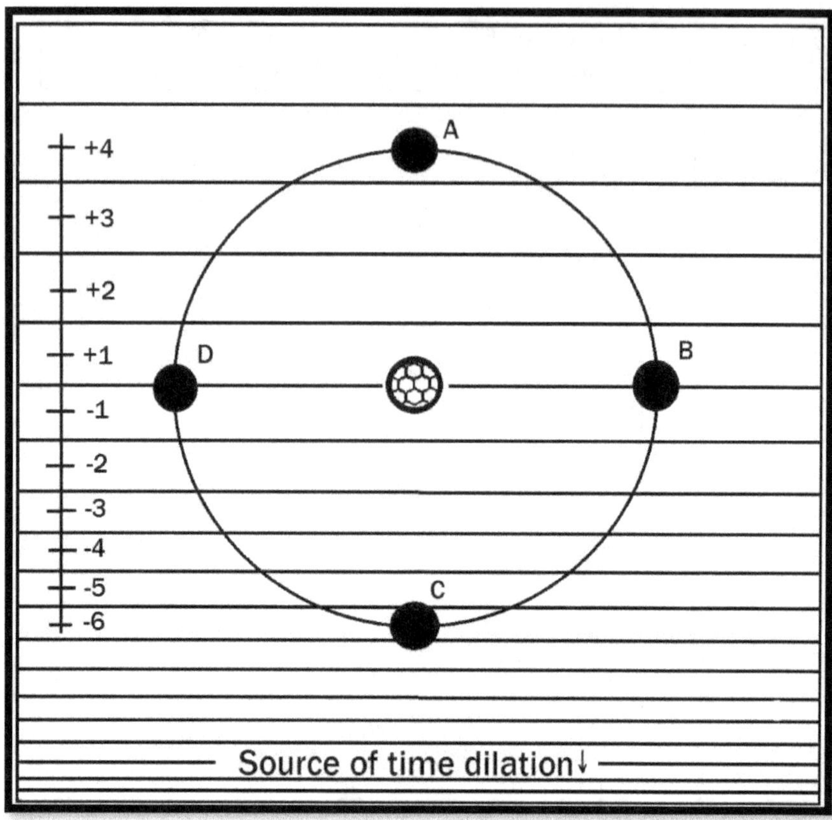

Illustration 3-3; Atom occupying relative space-time volume

This is an atom in relative space-time with time dilation nearest to the source having more compact lines of time. They equate to a greater spatial dimension in the atom in the slower rate of time. A slow rate of time seems to take on the effect of compressing space or giving an area more space. Space and time are proportional in that way. Slow time down and you have more space, speed time up and you have less space. The rate of time affects spatial dimension because they are two sides of the same coin of space-time.

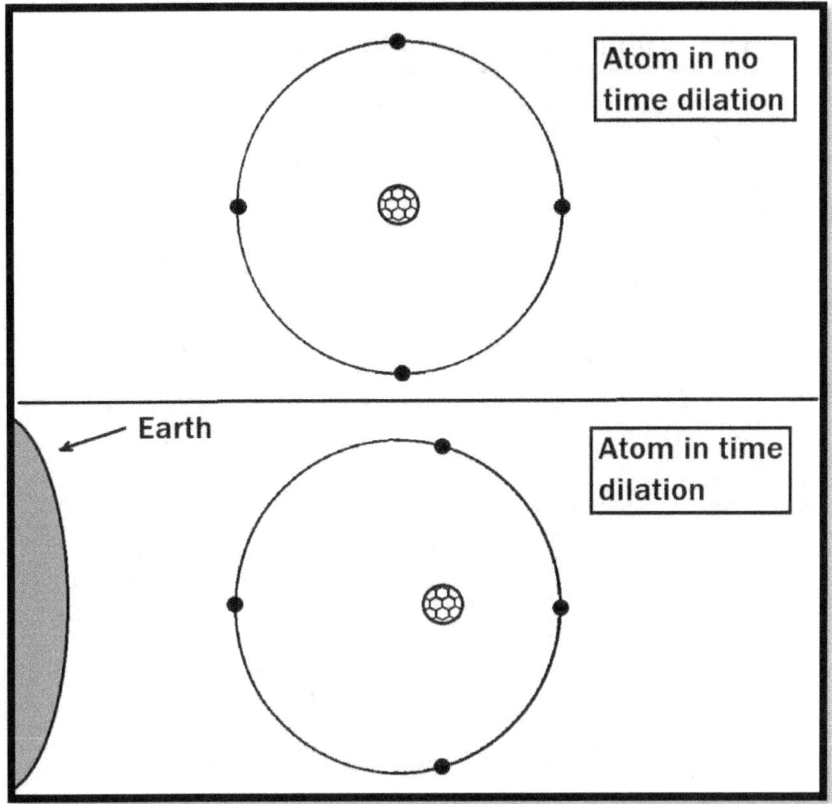

Illustration 3-4; Nucleus offset while in relative space-time

The top of this illustration shows an atom with no difference in time within it which would be considered flat or balanced space-time. Because there is no source of time dilation acting on it the nucleus is in true center. In the bottom half we add a source of time dilation which creates a field of relative space-time. That shows the change of characteristics in the atom. The effects of relative space-time change the nucleus's relative position. The closer distance in the field creates a higher energy potential leading to gravitation.

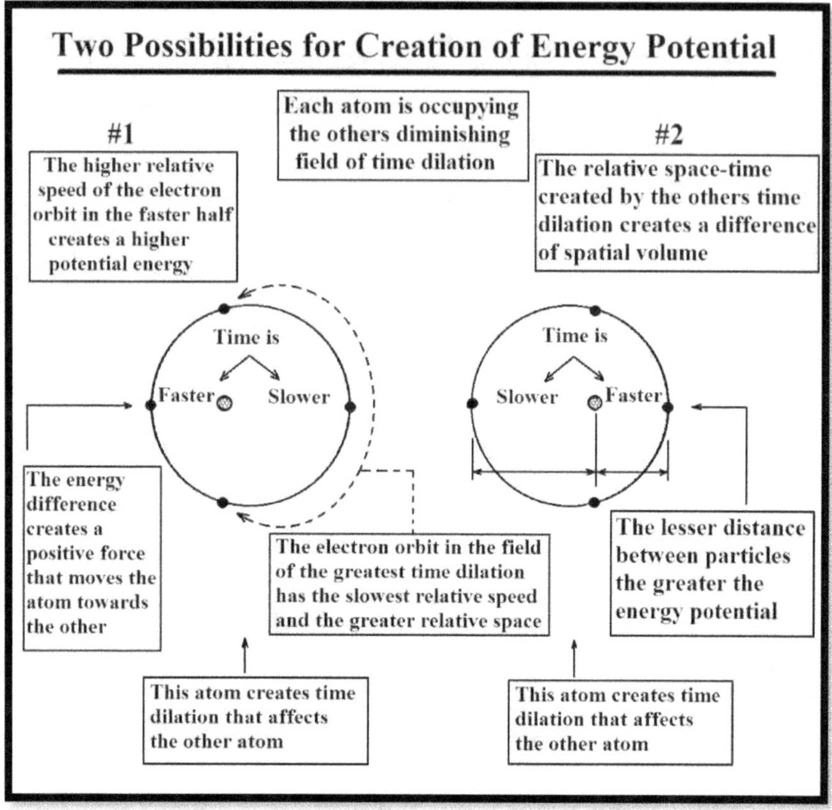

Illustration 3-5; Time dilation creates energy potential

Two atoms react to each other's time dilation to propel each towards the other. It is by means of time dilation and the relative space-time it creates. Time dilation is the only thing that exists that can create gravity and that occurs from the relative space-time that spans the distance. #1 is by reason of relative electron speed. #2 the offset nucleus signifies a difference in spatial dimension which creates the imbalance in the field of the atom. The imbalance creates the energy potential that leads to gravitation. #2 is primary in this book.

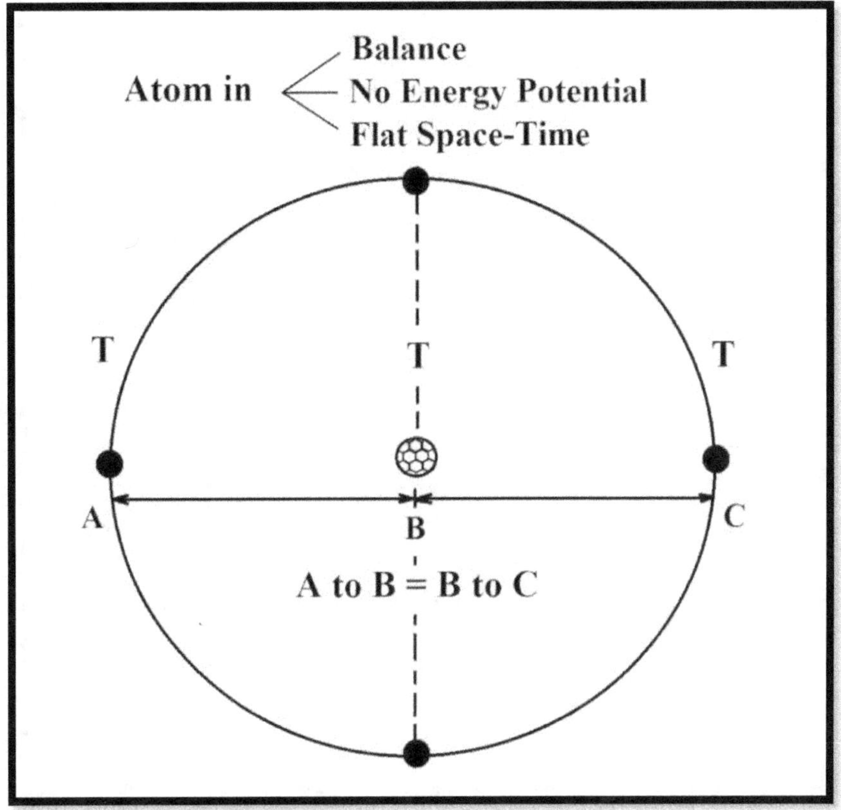

Illustration 3-6; Atom is balanced in flat space-time

This represents an atom in a balanced state and not being affected by relative space-time of time dilation by its differences. There is no difference in the rate of time or spatial dimension. In this configuration there is no difference of potential energy to convert to kinetic energy therefore this atom would remain stationary until the space-time it occupies changes to cause an imbalance in its field.

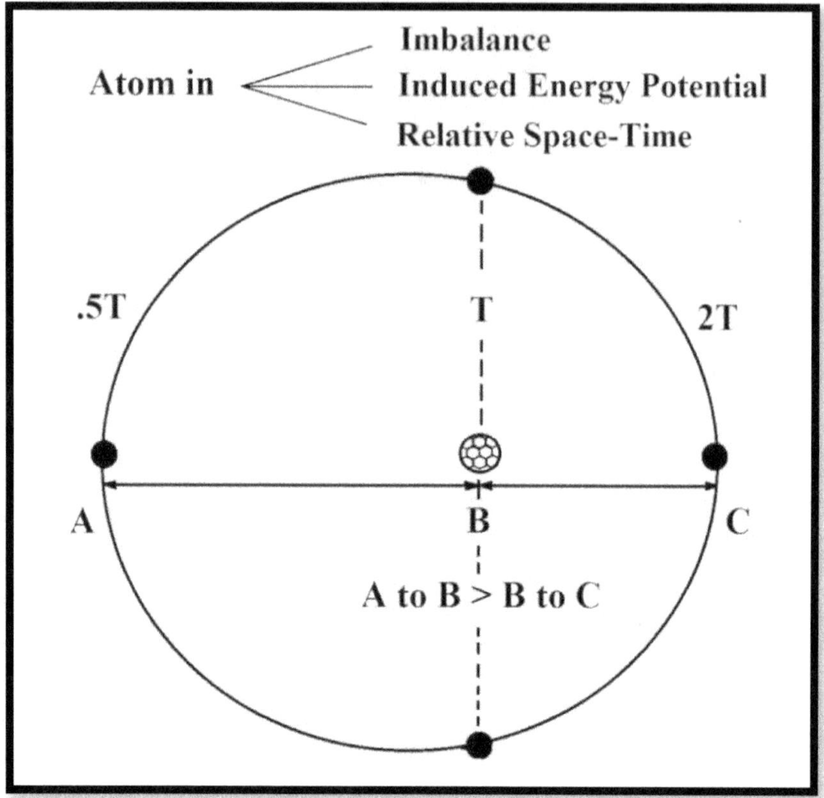

Illustration 3-7; Atom reacting to relative space-time

Source of gravity is on side A. An atom in relative space-time is an atom capable of creating gravitational propulsion within its fields. The spatial distance from A to B is greater than from B to C because time is slower in half A-B. When an area of space has slower time it takes light longer to pass through it and makes it larger. If it takes light longer to pass through than a corresponding area then the area it takes longer has more distance. That puts the nucleus off center which imbalances the field. The same effect works inside of the nucleus as well as the strong force field of the nucleons.

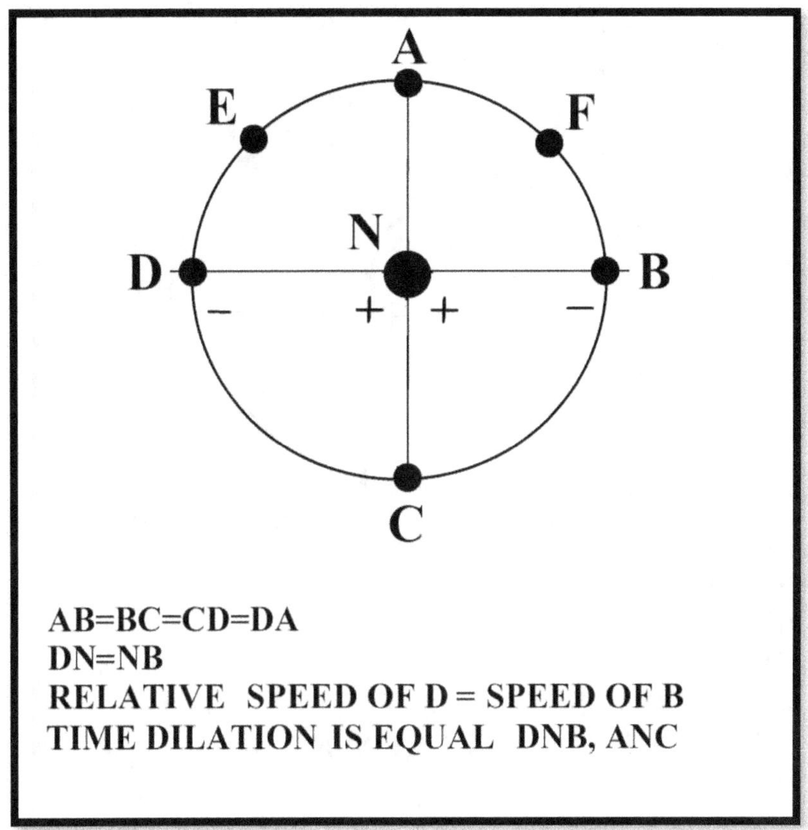

Illustration 3-8; Balanced electrostatic field in atom

There are no characteristics that create gravity or a higher energy potential in an atoms fields in flat space-time. Flat space-time is defined as having no gravitational source or differences of space-time that can create energy potential within the fields associated with atoms. The atoms fields are in complete balance, the rate of time is balanced in each opposing section which does not change the spatial volume leading to any portion of the fields having a higher potential. Flat space-time is considered balanced space-time.

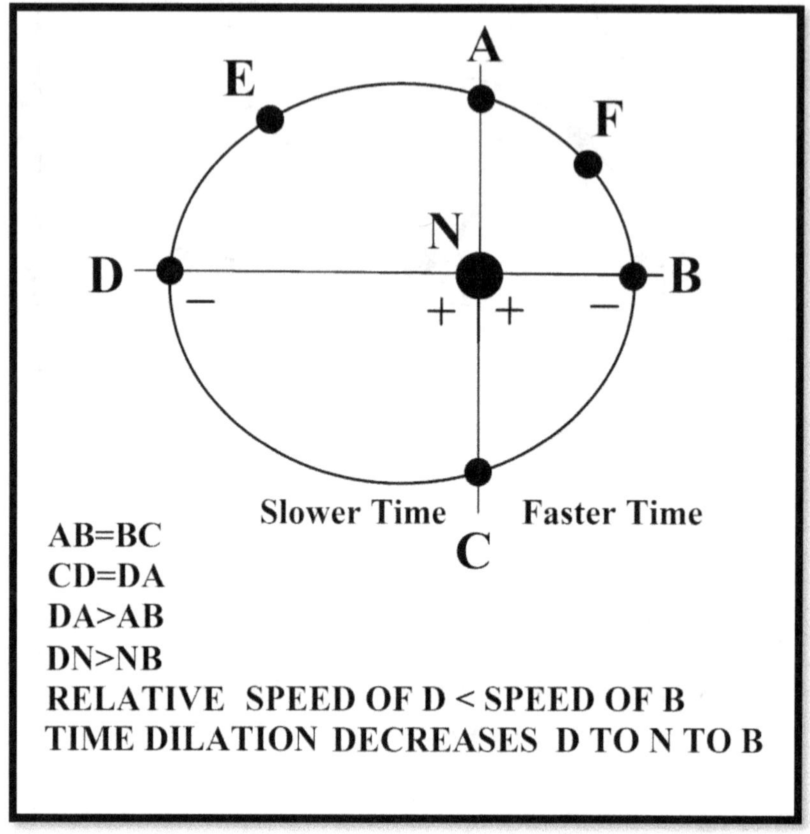

<u>Illustration 3-9; Imbalanced electrostatic field in atom</u>

This shows the effect that relative space-time has on a field of an atom. The lesser distance between the aspects of the field create a higher potential energy over the greater distance of the field in the slower rate of time. Any field within the confines of the atom has this characteristic such as inside of the nucleus or nucleons. Not pictured is the source of time dilation creating the relative space-time which is on the left side of point D.

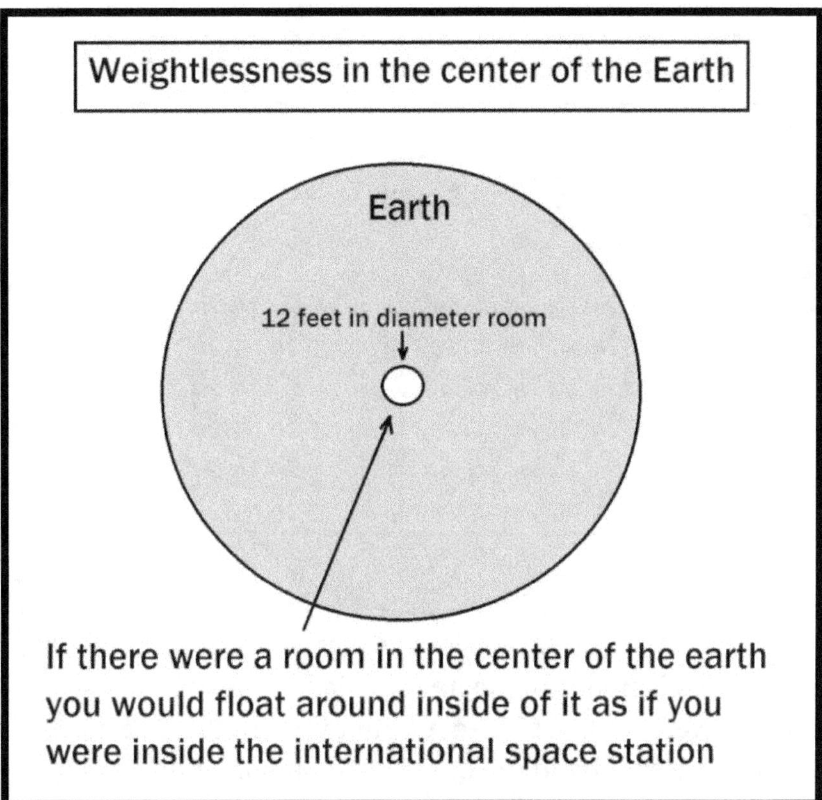

Illustration 3-10; Weightless in the center of the Earth

In the room in the center of the earth the force of energy potential of all directions cancels each other out creating a simulated neutral space-time. The time dilation of the earth is still there but there is no imbalance of space-time in the room. The rate of time and spatial dimension is the same everywhere in the room. The atoms in your body would have no deviation in the rate of time or spatial dimension to create a positive energy potential in them to propel you towards any side of the room, you would be weightless.

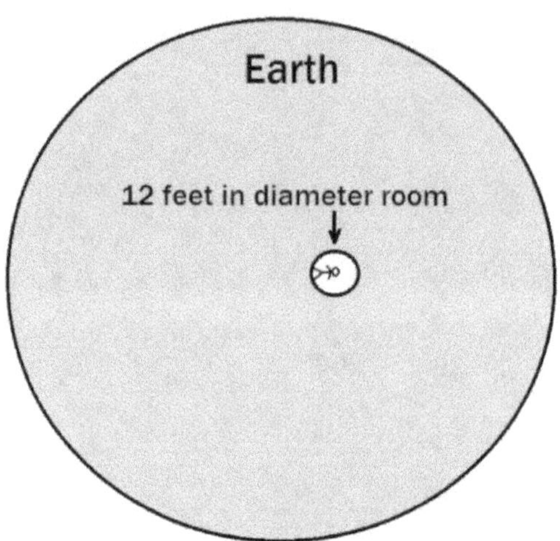

Illustration 3-11; Room offset from the center of the earth

This illustration shows a similarity of an offset nucleus in an atom in a field of relative space-time and direction of force in that atom and also in a room offset from the center of the earth. The offset would make the left side have a greater differential of space-time making that the floor. The similarity is that this would be how the atom would be pictured with an offset nucleus and the force on the atom would propel it towards the left side also.

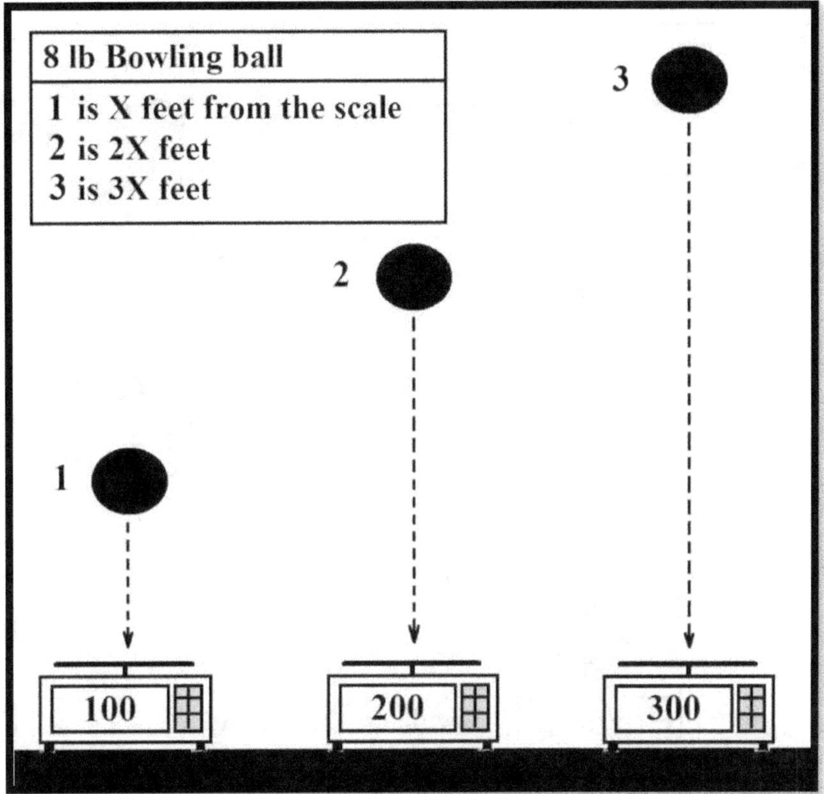

<u>Illustration 3-12; More force is added from gravity over time</u>

This illustration is designed to highlight the fact that the end result of gravity is more force upon impact. The higher the bowling ball is in height the greater the reading on the scale will be when it impacts. The point of this is to show that gravity is a force. The force of gravity is applied to the bowling ball and the longer it is applied the more force there is inherently in the ball when it hits the scale. Force cannot be created abstractly and then changed into a real force.

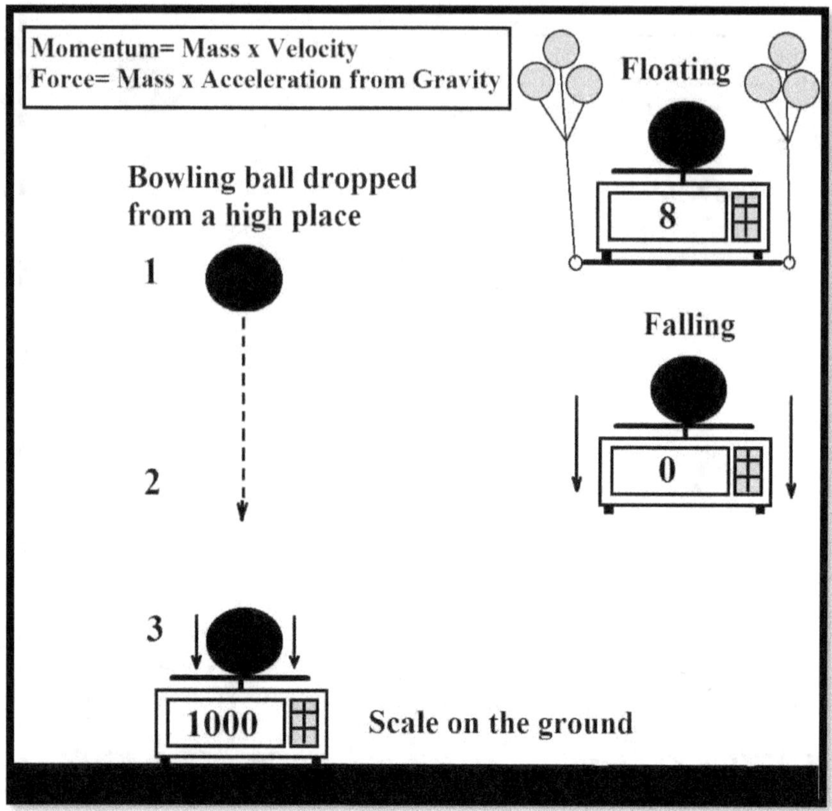

Illustration 3-13; Gravity is a real force and not abstract

The bowling ball falls at 32 ft. /sec², that is its acceleration from gravity. The longer it is falling the greater the velocity and the greater the momentum the greater the force that is released when it impacts the scale. If the scale wasn't destroyed it would read the total force absorbed in the ball during its free fall then settle down to read the at rest value of 8 lbs. The ball on the scale falling would read zero because there is no opposing force between the ball and scale, also when falling its absorbing the force (conserving) like a battery releasing it on impact.

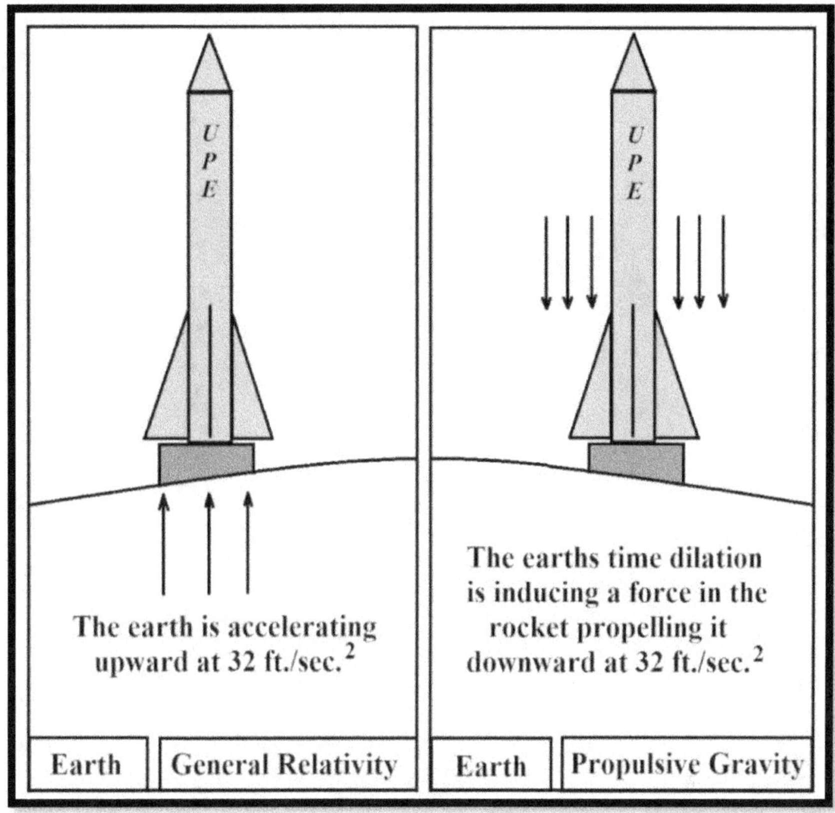

Illustration 3-14; The abstract nature of General relativity

This illustration makes a simple and basic comparison on how GR and propulsive gravity describe how gravity works for objects at rest on the surface of earth. GR postulates that the rocket is at rest and the earth is accelerating upward through space-time at 32 ft./sec.^{2}. Propulsive gravity states that the rocket has a force induced in it by the time dilation and relative space-time of the earth propelling it downward toward the earth at 32 ft. /sec.^{2}

How Time Dilation Bends Light

The bending of light is a concept commonly attributed to gravity. I say it is commonly attributed only because gravity is assumed to be the mechanism that is causing it to bend. There is a hint of truth to that philosophy but overall is not how the process works. Curved space-time is supposed to direct the light that passes through it and the curvature is then applied to the light. It is thought that the large mass of the sun bends space-time or curves it and the light follows a straight path through curved space-time. The description of calling space-time curved is not necessarily accurate. In a stream when water flows around a rock we don't call it curved water we call it an ebb or flow with changes of water direction and speed. Space-time being affected by a mass object and have created in it relative differences would be analogous to the stream. Changes in space-time would be induced by the mass object and create differences of the rate of time as well as spatial dimension.

How Time Dilation Creates Propulsive Gravity

Curved space-time of GR is based on time dilation and that is where we can agree creates the bending of light as well as gravity. But gravity is not bending the ray of light the two mechanisms share a common denominator but differ in how the effect comes about. Time dilation is the genesis of both mechanisms but they follow different steps in how they reach their end result. Using the term curved space-time is too broad and not specific enough to create the appropriate analogy for the mind to easily comprehend. If we place the emphasis on time dilation as the initiator and look at the basic characteristics of the diminishing effect with distance then we can perceive more clearly the basis of the process.

The effect that creates relative space-time is a diminishing one therefore any two points within space-time leading away from the mass object will have a differential of space-time. Since that differential is never ending and it is that which causes light to bend then light will always bend to a degree while passing through the universe.

***

The bending of light is called gravitational lensing because it is thought that gravity is the cause. If that happens not to be precise then this is another example that uses the wrong terminology. Since gravity is thought to be from curving space-time the analogy resembles the way light passes through a lens to change the lights direction. When we look out into the universe this effect creates a distorted view of the actual picture. In reality what we see as the universe is only an independent view from our perspective which is why our understanding of it is so limited. Gravitational lensing is a misnomer because gravity is not bending the light as defined by GR. There is a much more basic and reasonable explanation for the effect.

If that is the case then the concept no longer strengthens general relativity as the main factor for the phenomenon. GR would then be back to square one and only be an unproven theory. That is why this is an important section of this book. The theory provided in it may not be enough to change people's minds, there has to be along with it something that shakes the faith that people have in general relativity. Providing a rational more reasonable explanation as to why light bends passing a star would go a long way to give people a stepping stone. They may need that to help them take a step away from general relativity and towards a more rational theory of gravity such as the one I am proposing.

***

The bending of light was discovered by Sir Arthur Stanley Eddington in order to prove Einstein's theory of general relativity. Eddington traveled to a location that was due to have a full solar eclipse. He used his telescope with a camera mount and took successive photos of the eclipse from beginning to end to find out if light of a star behind the sun was being bent and to prove it with photographic evidence. After he had gotten back to his laboratory it took some time to evaluate the photos but in time he contacted Einstein with the good news which helped cement the theory of general relativity into the history books. It was all over the newspapers and Einstein became a household name. That theory has held dominance in the physics world ever since and to this day the concept of gravitational lensing helps keep general relativity as the prime concept for defining gravity. Considering we usually don't get things right on the first try then this is something we should look at more clearly in order to reevaluate. To look at what it actually tells us and equally what it doesn't.

What if a better explanation comes into existence to describe why light bends? I'm not suggesting a totally new and different mechanism only a similar one based on the same factor with only a redefining of that factor or factors. General relativity has reigned for over a century and considering how little we've come to building gravity powered ships I'd say we dropped the ball. We've had a century to make progress but the general consensus was to steadfastly hold on to one theory over all others. To follow something blindly in an unknown direction assuming it was flawless. GR should have initiated an enthusiasm and thirst for knowledge in people who experienced such a mind bending new concept. Instead it held them back to become followers because they thought that was the end of the journey.

***

This theory doesn't use gravity as the means of light bending it uses time dilation which is a level deeper than gravity. Yes time dilation creates gravity and yes it also bends light but gravity does not bend light. Even though GR's curved space-time is close it does not clearly define the exact process. I would say that GR's explanation of why light bends is closer to reality than the mechanism of gravity it proposes. That is consistent with my statement that this theory of propulsive gravity shares the factors of time dilation and relative space-time but diverges after that.

There are two different mechanisms with two different results that stem from time dilation. More specifically by how time dilation affects space-time by creating relative space-time between two points. I understand that in general relativity the concept of curved space-time comes from time dilation and proposes that it creates gravity but it also proposes the same mechanism bends light. I say that both gravity and the effect of light bending share two things only, time dilation and relative space-time. They differ in how each result manifests because they have a different process after those common factors.

I may be simply following the same reasoning for light bending as GR and just nit picking its explanation but there is no real explanation given other than light follows a straight path through curved space-time. I go a little further and explain the exact process specifically. Besides that fact GR was proven only as far as the bending of light was proven and that really only proves the effect of time dilation. As far as gravity goes there is no real proof that it also creates gravity in the context of how GR is presented.

The real process is very simple, when light passes through time dilated space-time it bends depending on the strength of the time dilation. Time dilated space-time has layers of time that pass slower than the next layer and faster than the previous layer emanating around any object of mass like layers of an onion. Since the speed of light is always at the maximum speed limit it's always on full throttle passing through the universe. When it passes through an area where time is affected by a mass object the differing rates of time channel the light as if passing through a lens. The relative space-time is that lens and the differences of the rate of time is the mechanism not that of space.

The key to the mechanism is that light is always at maximum speed. The mechanism is that light will travel farther in the faster rate of time and lesser in the slower rate of time. Since relative space-time is defined as having those differences then that alters the path of light. The light ray occupies spatial dimension and the difference in the rate of time spanning it causes a portion of the light ray in faster time to travel further. The portion of the light ray in slower time essentially drags or lags behind causing the light ray to arc according to the differential of the space-time. The amount of the differential of space-time is determined by the mass of the source object and the distance from it. Since both gravity and the bending of light share time dilation and relative space-time they both follow an inverse square function. Because they both follow an inverse square function then light is assumed to bend due to gravity even though that is not the case.

***

Here is an example for you to imagine, when light is bending as it is passing a star it's bending because time is changing the distance that parts of the light ray are able to travel. Imagine the light ray as a cylindrical tube traveling through space. Its traveling as fast as space-time allows but when it's passing a star the side of the tube facing the star is in slower time with the opposite side in faster time. Since it's already at maximum speed the difference in the speed of time changes the distance the tube "light" can travel forward. So what you will see is a dragging of the tube on the side facing the star which is the side of the slower rate of time. That side will not be travelling as far as the side in faster time. When you combine those characteristics of the two sides with differing dimensions, one shorter and one longer the only way to match them up into a single ray or tube would be to combine then into an arc.

Another example is to imagine that a ray of light is a car with four wheels and on each side the wheels diameter is determined by the speed of time in the time dilated space-time. Faster time would make the wheels on one side a larger diameter and slower time would make them a smaller diameter. The reason for that is because in faster time light would go further than in slower time so it would have a larger circumference. The larger wheel would go further than the smaller wheel with each revolution. All of the wheels on the car turn in unison as though the wheels are welded to the axles and the steering wheel has a fixed position. The speed of time is also determined by the time dilation in an area of space-time that the light is traveling through because it creates the relative space-time. That would make the diameter of the wheels on the car variable and the rate of time would make them vary. Time dilation would essentially direct the car left or right according to which side the time dilation was affecting.

By changing the diameter of the wheels on one side making them smaller that would cause the cars direction to veer towards that side. That side would be occupying the space-time of the slowest rate of time. The slowest time would be the side of the car facing the star and the direction would be controlled only by variances in the speed or rate of time. The time dilation can be of any strength not just extreme time dilation near a star or black-hole. If you imagine that you could see the differences of time in the universe you would also see light bending and curving as though it were a car driving through an obstacle course. That may be the closest analogy to curved space-time in GR but it also defines that the process is related to relative time and not curved space-time which is another terminology error.

One other example to ponder is to look at a set of railroad tracks that form a half circle to resemble an arcing trajectory. Measure each rail of that track with the inner rail representing the side of the ray of light closest to the star and the outer rail the farthest side. The inner rail would of course be shorter than the outer rail representing the light rays varying travel distances due to the varying speeds of time caused by time dilation and relative space-time. The light ray as a whole would have half of the ray traveling farther than the other half because the half in slower time is being limited in travel distance resulting in the bending of the light ray.

***

Time dilation and its effects diminish with the distance away from any mass object as does the arcing effect on light. If you measure the effect of light bending for a specific distance or duration it will follow the inverse square law for any point from a mass. But since light bends in parallel time dilation and the distance it travels increases with distance from a mass the overall bending doesn't follow the inverse square law. The farther away from the center of a mass you measure the greater the area of parallel time dilation exists for light to pass through. If you imagine the center of a star and then draw two lines from that point at sixty degrees you make two sides of a triangle. If you measure the distance between the lines it gets greater the further you measure from the center of the star. That area between the lines represents the most parallel time dilation. As the distance increases then so does the compounding effect of the time dilation on the bending of light.

In the illustration at the end of this section this is visually explained. Comparing the area of parallel time dilation at ten radiuses to two radiuses from a star and you see that the light has five times the duration of travel time for it. If time dilation follows the same equation as the force of gravity then the light ray A is passing through twenty-five times weaker time dilation than light ray B. Light ray A has five times the duration of parallel time dilation to pass through. With its compounded effect that gives it an overall bending that is only five times that of light ray B not twenty-five. So even though light is passing through space-time far away from a star it is likely bending much more than previously imagined.

In the very least if you can agree that it is very likely that time dilation which creates relative space-time bends light then you are also agreeing that light doesn't really follow a straight path through curved space-time and even though they may seem similar the distinction should be made. The term curved space-time and differential space-time may be the same thing technically and create gravity along with the bending of light. But those are two different things with two different mechanisms as to how they come about.

Gravity does not bend light because there is no mechanism for that to happen in the context of gravity. You may be against my line of reasoning which is ok but I am following the path that is put forth by the terminology used to describe the characteristic. The inadequate terminology is really a characteristic of being vague, unspecific and incorrect. Just because light bending occurs within a gravitational field doesn't mean gravity is causing the light to bend. It only means that a gravitational field is a field of relative time dilated space-time.

***

The point of separation of the two effects of gravity and light bending is that gravity is the process that occurs for objects of mass. Objects of mass are structures of fields that have opposing polarities. Light is not a structure and does not have opposing polarities it is a single polarity and not a field. Therefore the mechanism of gravity is not affecting light causing it to bend. They are two different things that share a similar foundation.

If we change our way of thinking about these things then the curved space analogy and the definition that gravity is not a force then the foundation of GR is being shaken. What we have held on to for so long for the last century may be on its last legs. GR has reigned for so long people may not be able to accept any other possibility for the creation of gravity. That is a shame since that would be limiting mankind's knowledge and ability. By doing that it's going to take us longer to unlock the mysteries and get out there into the universe to explore.

If the theory of gravitational lensing is altered, not rendered obsolete but just altered that sends us back to 1919 when Sir Arthur Stanley Eddington and his team found proof that light bends as it passes the sun. That is all he did, is prove that light bends which proves the characteristic of relative space-time due to time dilation. So his findings should not have given GR full validity and by virtue also proved how gravity functions. Even though Einstein predicted that light bends one single prediction coming true should not have been a qualification to accept GR fully.

To this day the bending of light being assumed is due to gravity only because its assumed that curved space-time creates gravity is the only thing that supports GR. Everything that supports GR is an assumption based on its only accurate prediction. Everything used as proof can be tied to the effects of time dilation and relative space-time but that does not support the entire scope of GR. Even though it may seem lights path acts like a straight line and curve along a path in relative space-time it is not gravity doing the curving so only supports the effect of time dilation. But like I said before they only assumed how light was bending because it was preconceived. They proved it was bending but they didn't prove how exactly. If you can agree that relative space-time as a consequence of time dilation is responsible for both gravity and light bending and are able to see how the two mechanisms differ then there just may be hope after all to move forward.

Light bending *"Is simply the consequence of an altering of trajectory as a result of time dragging"*

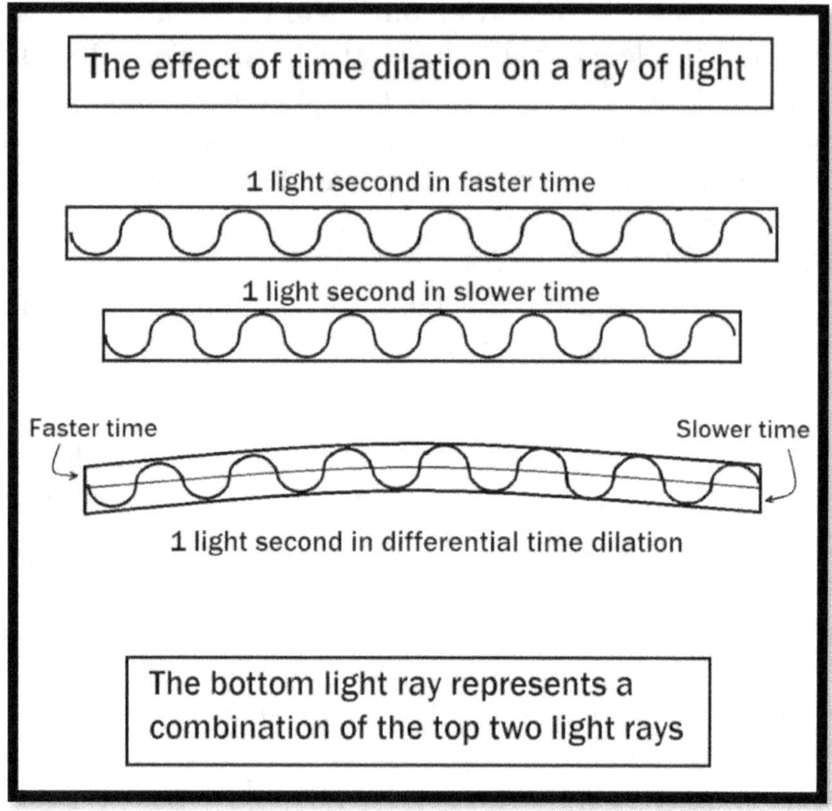

<u>Illustration 4-1; The effect of time lagging on light</u>

This shows a ray of light sliced down the center, where one half of the ray closest to the star and traveling parallel to it is in slower time than the other half. That means the half in faster time is going further and the half in slower time lagging behind not going as far (time lagging). This relative difference causes the effect of an arcing to occur in the light ray. This is why light bends as it travels through parallel relative space-time. Differential means there are differences of space-time that create the effect and lead to the end result.

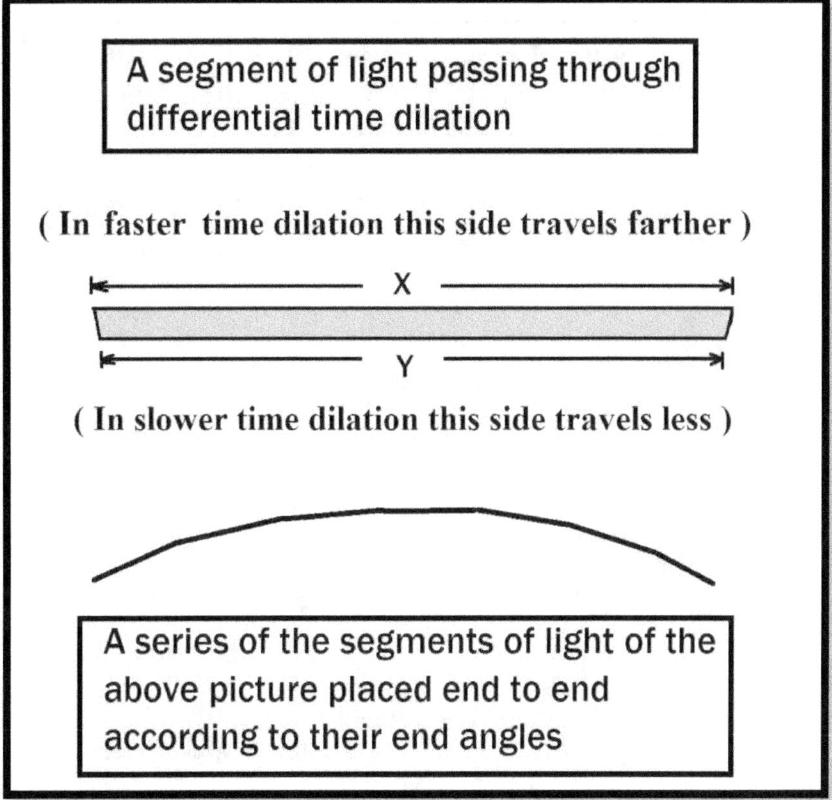

Illustration 4-2; The geometry of time lagging on light

This is another representation of the previous illustration. It shows another way to look at the phenomenon since light exists as a three dimensional ray and the difference in time caused by relative space-time will specifically affect its trajectory. If you look at the light in segments then align those segments end to end according to their end angles they will not produce a straight line. That is an analogy of the actual circumstances, it causes a smooth curve not flat segments end to end. The analogy is an aide to help in understanding. (Differential means there are differences)

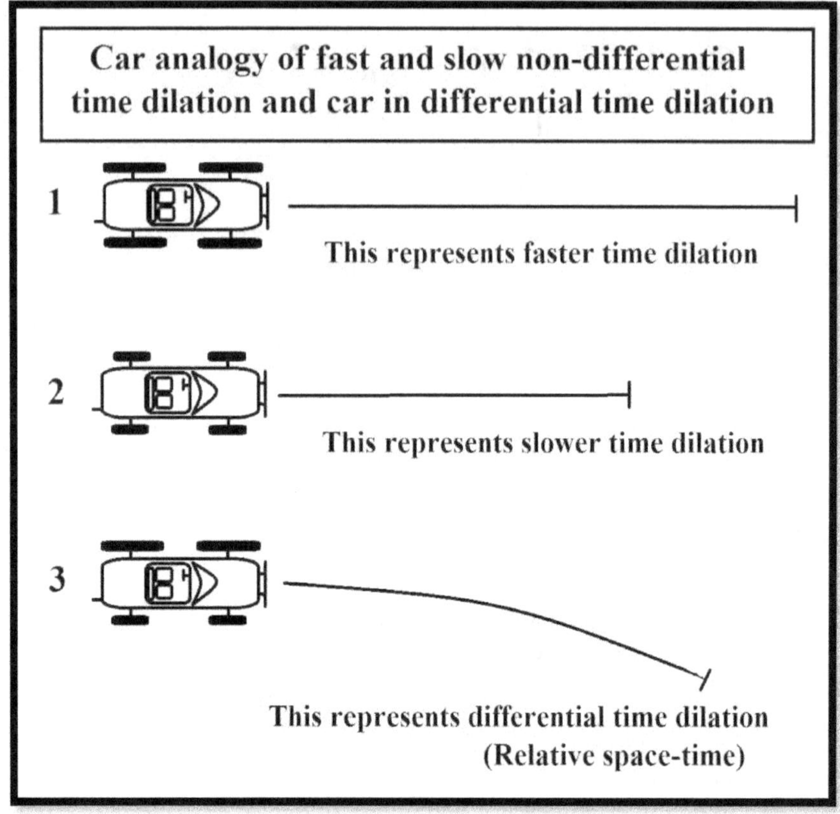

Illustration 4-3; Car analogy of why light bends

The cars axles have a set speed of 60 rpm which represents the set speed of light. Car 1 has 6ft circumference wheels and car 2 has 3 ft. Because light travels farther in faster time car 1 will go farther with larger wheels than car 2. If a car were to travel through relative or differential space-time passing a star which means time is slower at the surface and gradually speeds up with altitude it would represent car 3. One side of the car would be in faster time and the other in slower time, the car would arc towards the slower time.

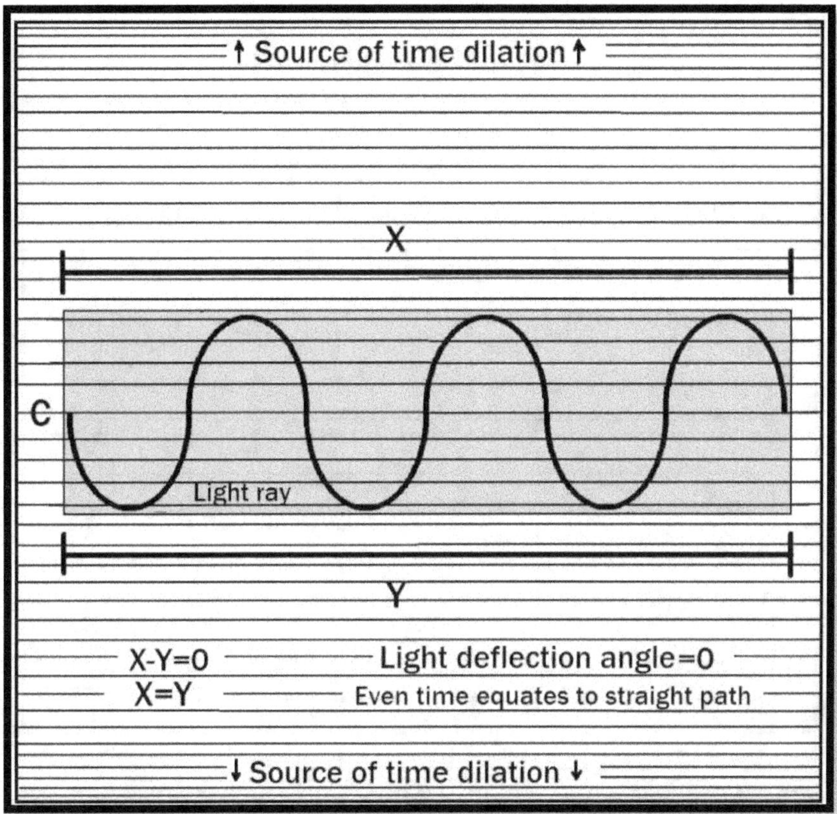

Illustration 4-4; Balanced space-time effect on light

This example shows a ray of light passing exactly between two sources of time dilation so that the space-time it's passing through has no deviation in the rate of time. It shows each side of the light ray is the same therefore there would be no deflection angle. This would also be an example of a light ray passing through flat space-time but in this case it is neutral space-time as a result of being balanced between two sources.

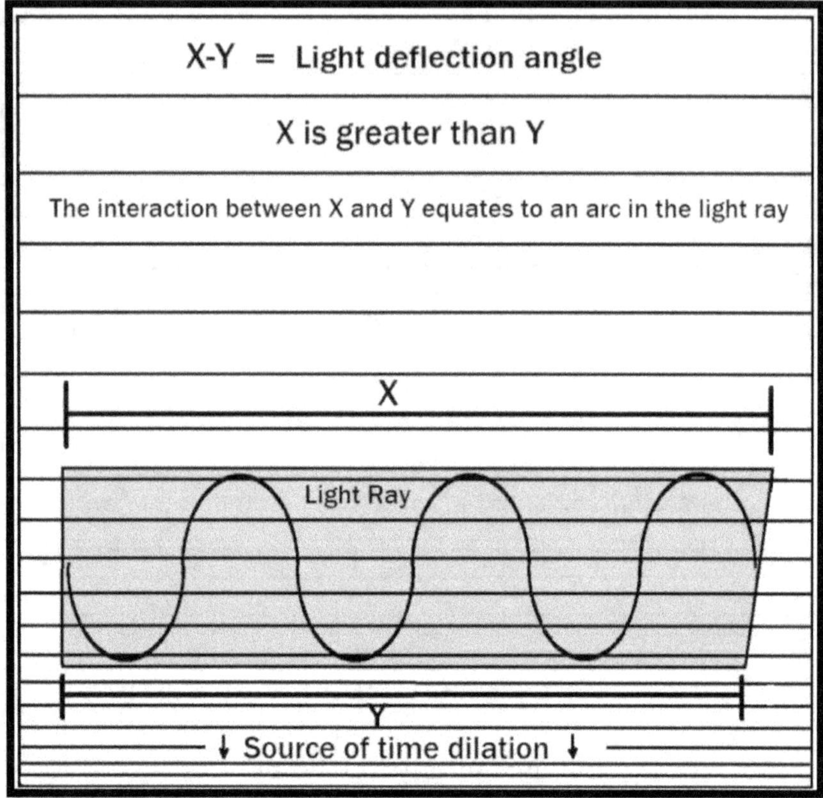

Illustration 4-5: Relative space-time effect on light

In this example there is just one source of time dilation acting on the light ray. One side of the light ray would be longer than the other because that side is occupying faster time. The end result creates an arcing trajectory. Time dilation causes a light ray to have varying lengths due to relative space-time. One side would be longer and one side would be shorter and to compensate the light ray travels in an arcing trajectory.

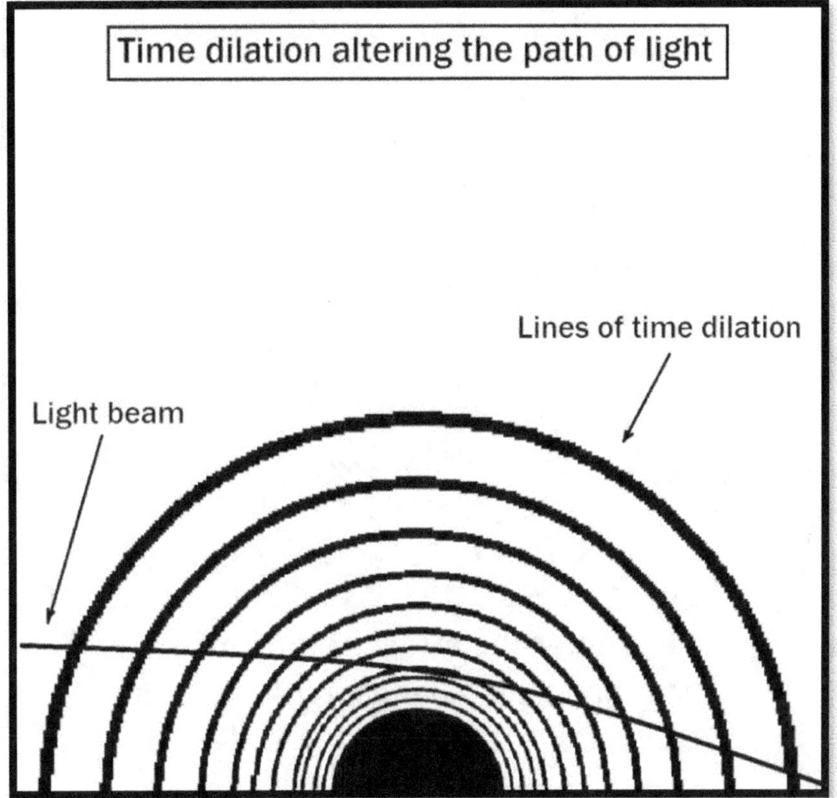

<u>Illustration 4-6; Light ray through relative space-time</u>

This is a simple illustration of a star and the time dilation around it with a light ray passing. The time dilation is the strongest nearest to the star. The light ray would bend the greatest nearest to the star. The reason for that is the time dilation will diminish according to the inverse square law equation from the center outward. It would be an exponential decline of time dilation. Even though it declines with distance away from it any relative space-time will alter the trajectory of a light ray if there is a differential of time.

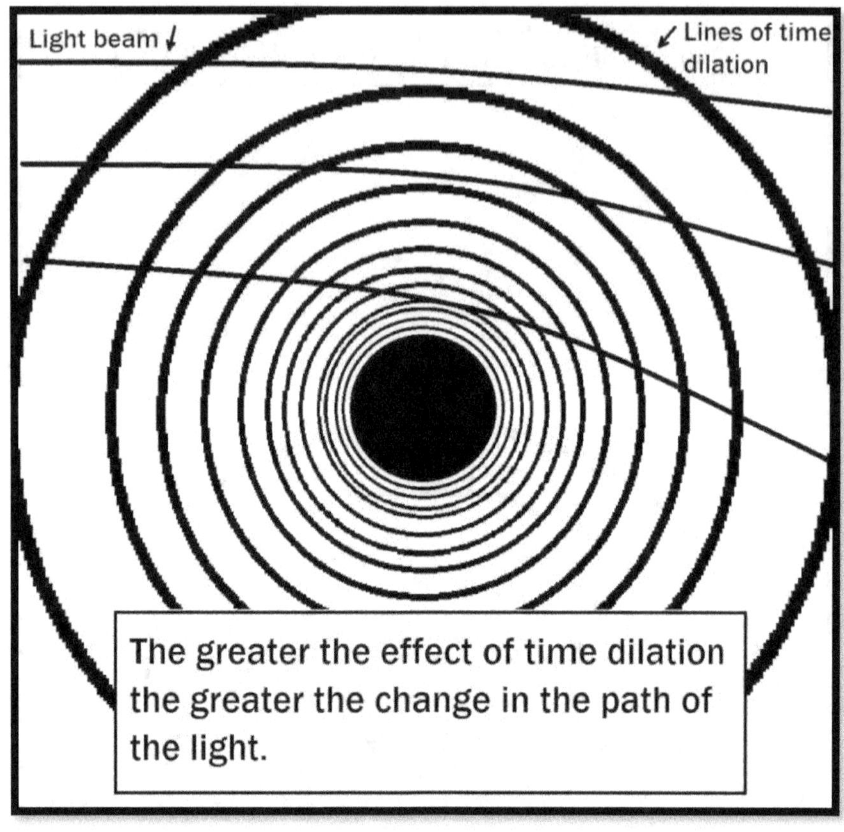

<u>Illustration 4-7; Multiple light rays in relative space-time</u>

This illustration shows three light rays in three different strengths of time dilation surrounding the star. The light ray closest to the star bends the most and the next light ray bends slightly less and the next light ray bends even more slightly less. According to the inverse square law equation the time dilation effect around the star never reaches zero no matter how far away. That means no matter how far away a light ray passes by a star it is always bending from some significant to insignificant amount.

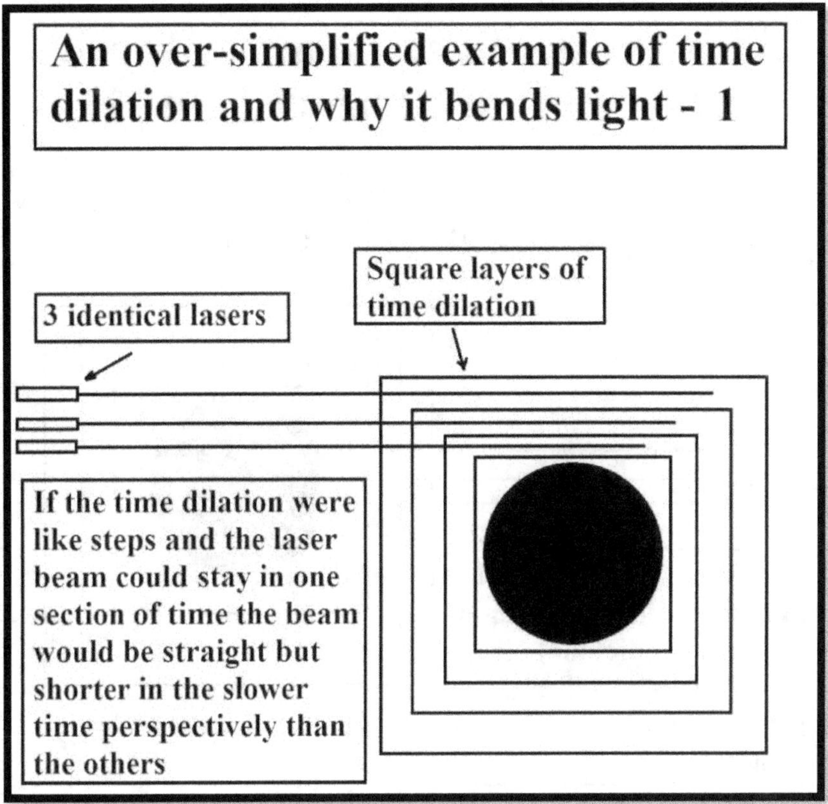

Illustration 4-8; Over-simplified example of light bending 1

These next four illustrations are over simplified to be more specific about how time dilation bends light. By simply making the time dilation square that makes it easier to comprehend the process of light bending as it passes through it. Each section of time dilation is stepped and inside each step time is one rate with no differential, its slowest nearer to the star and faster each step away from the star. The laser light therefore travels the least far closest to the star and further in the next two steps.

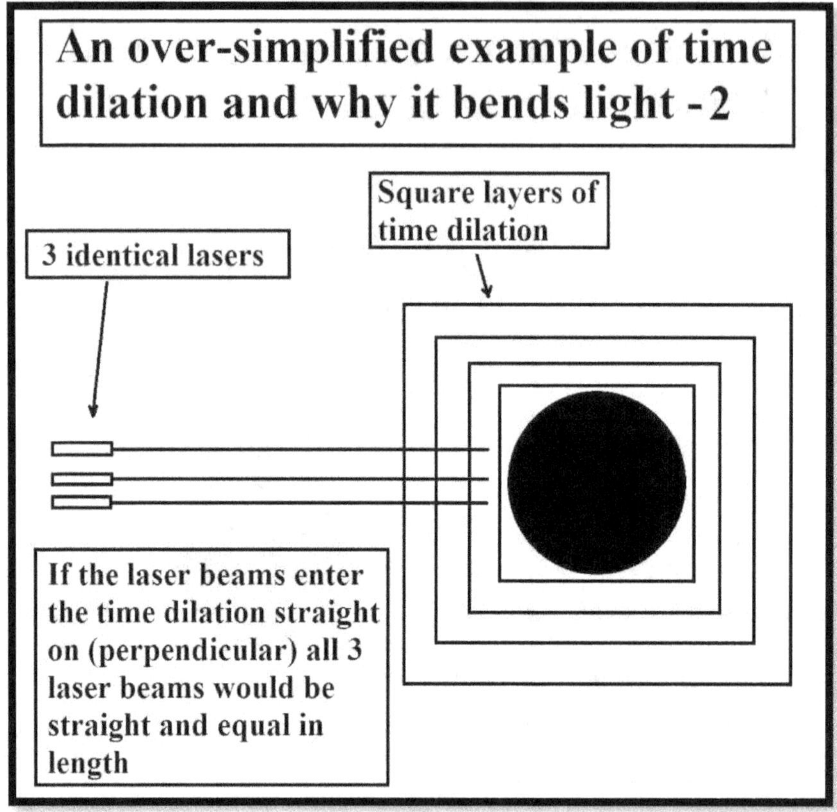

Illustration 4-9; Over-simplified example of light bending 2

If the lasers were set up to aim directly at the star they would be penetrating the time dilation at a right angle and all three would be affected the same. They would enter each layer of time dilation in unison and be affected simultaneously. The bending of light only occurs when passing through the time dilation parallel to it because the ray occupies three dimensions of space-time and being parallel creates effects of the uneven or relative time differences associated with relative space-time.

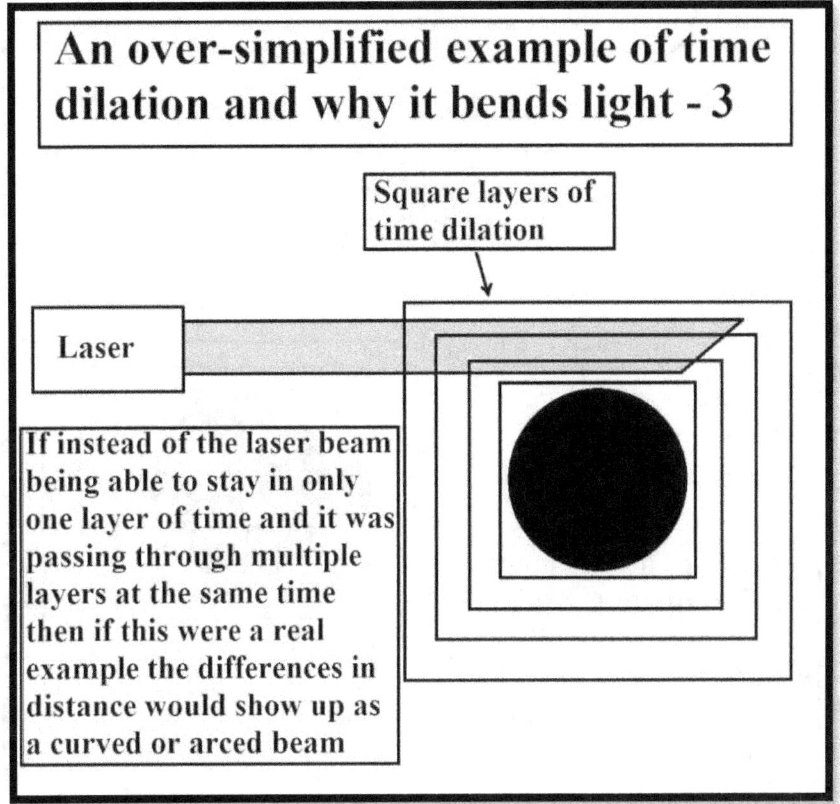

Illustration 4-10; Over-simplified example of light bending 3

The light will bend when it is passing through the time dilation parallel to it because that way has the light ray occupying multiple rates of time. In each successive layer of time the light ray will only traverse so far as time allows since light is at a maximum speed already. As pictured above the end is drawn as an angle for simplification. In actuality the end isn't angled it is the entire ray of light that has its trajectory angled making an arc instead.

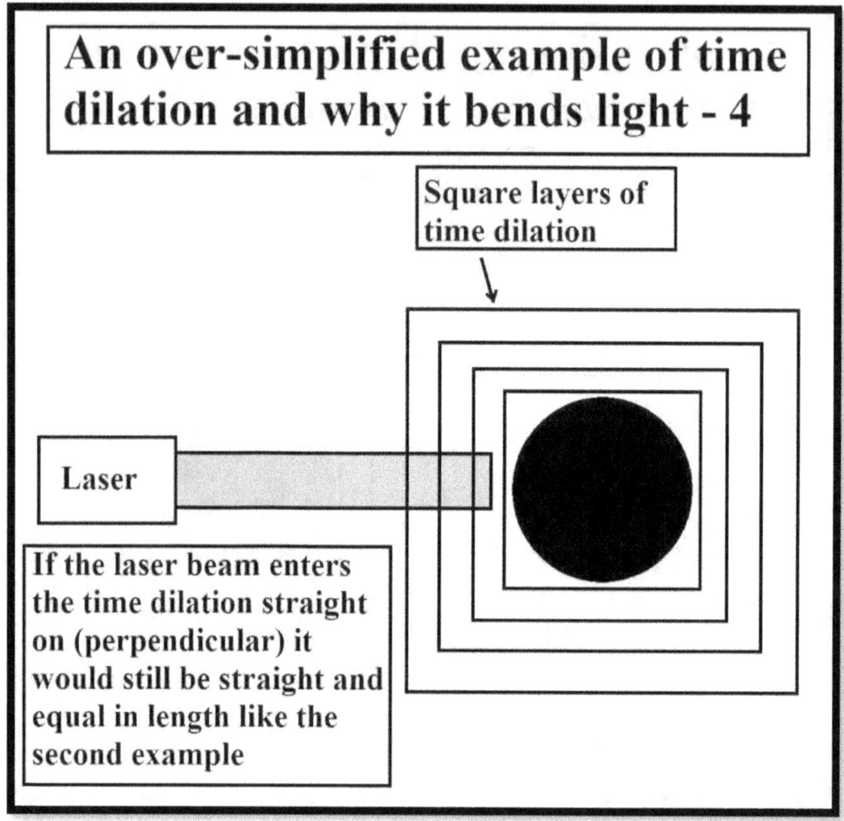

Illustration 4-11; Over-simplified example of light bending 4

If the three lasers are replaced with one large laser beam the original result would still apply. These examples are designed to show the bending occurs only when light passes parallel to differential time dilation (relative space-time). If you slice a light ray down the center one half would be in faster time and the other half in slower time, with the nature of light having a limited set speed that characteristic is most likely the reason light arcs while passing parallel through lines of time dilation (relative space-time).

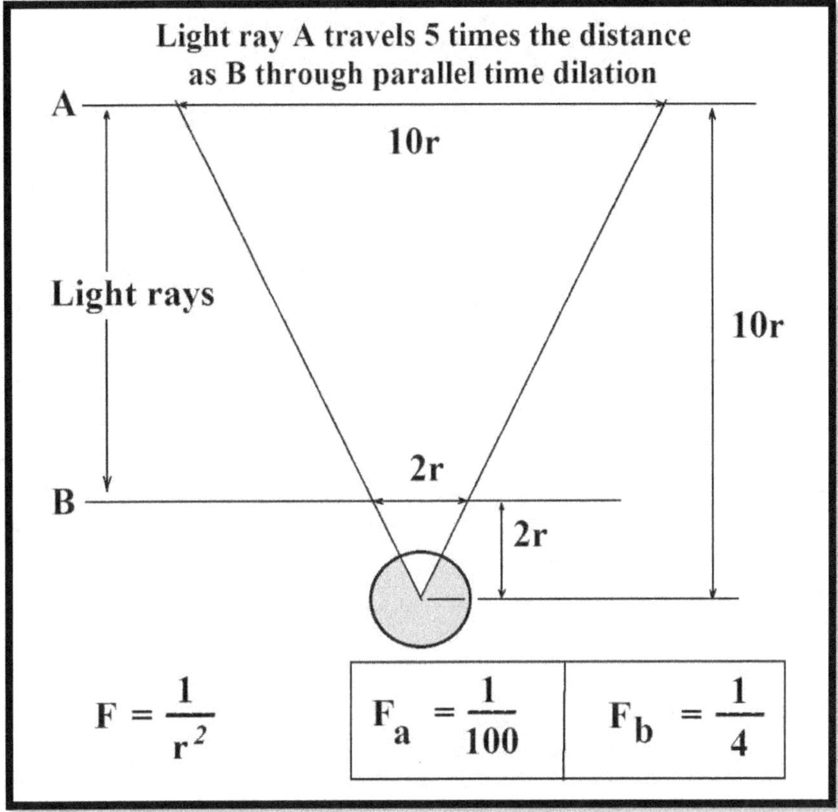

Illustration 4-12; Parallel time dilation effect compounds

This shows that with distance from the sun the amount of parallel time dilation in relative space-time increases. In this example it increases five times. That means that even though the time dilation diminishes for light ray A it stays in it five times the span or five times longer. The bending amount of light doesn't follow the inverse square law because it has more parallel time dilation to pass through. The bending of light compounds so with the longer distance it bends more than the 2r distance. It bends as the 2r x 5 which equals 10r.

How Time Dilation Creates Propulsive Gravity

Artificial Gravity Propulsion

The theory of gravity in this book is laid out in detail of exactly how the process of gravity works. In defining the exact process we can then use an analogous mechanism to simulate propulsive gravity. We can then utilize it to create the next generation of ships to make space travel commonplace and safer. It seems so elementary now that gravity is a direct result of the time dilation and relative space-time an object is occupying. Even though GR had a similar view it could not utilize the concept to simulate gravity as this theory does. This is a big step forward in a space faring society and all it took was to define gravity more precisely.

The key to gravity is energy potential induced within a field which also makes it the key to simulating the propulsive effects of gravity. There are two potential paths we can take to simulate the mechanism of gravity. They are based on reproducing the characteristics created by time dilation which are the diminishing properties of relative space-time. The first and most direct and likely the most difficult is the production of artificial time dilation to manipulate the space-time a craft would occupy. The second is reproducing the effect of relative space-time which creates the imbalance in an atoms field or fields. That centers on the creation of potential energy and creating a difference of potential within the electromagnetic and strong force fields constituting the atom and within the nucleus. Since gravity is a force of this nature it may be the best place to start. We have a destination and a starting point and now we must navigate what is in between.

There is a large gap in our knowledge and understanding of space-time. It is so fundamental yet the hardest to conceive of because it's the basis of our existence. It is the fabric of the universe and we barely made a dent in understanding that. Electromagnetism on the other hand we have a much firmer grasp on its understanding so we're not starting from scratch. If we are to find a safer more environmentally friendly source of propulsion we have to start somewhere and without even knowing we have already started in fact Nikola Tesla may have already started.

***

At this point in time being at the starting point an outline of the process to build a technology like this is unlikely to be accurate. I don't know the exact specifics on how to build a machine to simulate time dilation. I haven't tried to build anything yet, so far it's been a project in the mind more than anything else but this is a place to start. This concept of gravity is new and has no precedence so this is the beginning of a fascinating journey. Who knows if I will make the attempt but I'm sure that if we put our minds to work we may just surprise ourselves. I will present an outline on what I imagine the propulsion design may be and how to possibly achieve it or at least begin. It may help in the designing of it to imagine what the final product will end up being. I will cover all possibilities that I can imagine associated with this type of technology. Maybe someone out there with an inventive mind will already possess a technique that may lead the way towards accomplishing this task. It may be in their mind with no feasible use and learning of this possibility will open the door and they will put two and two together and have a serendipitous moment.

***

Propelling machines around our environment occurs in a few basic configurations. On a bicycle you have to apply a force downward on the pedals which in turn transfers that force to the back wheel via the drive chain. The force applied to the wheel spins it around propelling the bicycle. On a car the design is similar but the force comes from the gasoline being burned inside of the internal combustion engine. It's more complex because it has greater power and needs to have a transmission between the engine and the delivery of the force to the back wheels to propel the vehicle. On an airplane which also has an engine delivering the force to propel it is more direct and requires less of a mechanism. The rotating engine can apply force directly to the air to propel it by using a propeller. By rotating the propeller it pushes the air behind the plane which essentially pulls the plane forward. A boat or submarine is similar but pushes the water behind it pushing the vessel forward.

The reasoning behind explaining those different types of propulsion mechanisms is to highlight the multiple ways craft can function to complete the same basic task of motivation for the purpose of transportation. Some have simple ways to apply force and some have complex ways in the overall design to motivate but the end result is the same.

Space travel with our current form of rocket propulsion consists of burning an explosive fuel in a combustion chamber. It directs the exhaust out of a nozzle creating a high pressure force out of the bottom of the rocket. That method pushes the rocket forward in order to propel it. The energy created with the combustion of the fuel is specifically directed in the opposite direction that the rocket needs to travel. The force that motivates the rocket is applied to the area of the rocket opposite of its direction. That is a basic rule to motivate any device to move in any medium. If you apply or induce a force on one side of an object or machine it moves it in the direction of the opposing side. With a higher potential on one side and a lower potential on the other the movement is always from higher to lower.

Propelling a ship using gravity propulsion through space-time is no exception to that rule. A force needs to be applied to the craft on the opposite side that the craft is desired to move. It just so happens that that is exactly how gravity moves an object through space-time. The relative space-time that an object is occupying creates an imbalance of time and space inside of the structure of its atoms. That imbalance creates a potential difference in an atom with the higher potential being on the side opposite of the atoms motivated direction. Since this motion was induced by relative space-time via time dilation the object moves by its self-propulsion toward the source creating the time dilation.

The first possibility and most difficult surrounding this propulsion method is being able to create artificial time dilation. The type of time dilation can either be increased or decreased time dilation. That means to have the ability to decrease or increase the rate of time for the act of lifting the ship up off of the ground or altering its trajectory. The time dilation mechanism will have to create a slower rate of time above the ship and or faster rate of time below the ship. It may be a process that has an inverse function automatically such as if it slows down time above it then time speeds up below it and vice versa. It would also be important to be able to direct the field being created towards any direction. The process would be useless if there was no way to control which direction the ship moves.

When I say creating slow or fast time above or below I'm referring to a bubble of artificially created time dilation in the space-time surrounding the ship. The slow or fast time would be above and or below and gradually dissipating towards the other side of the bubble of space-time. It would have to simulate natural time dilation which I consider differential time but usually refer to it as relative space-time and anything that occupies its space has differential time from one point to another. If you know how this process of gravity functions for one atom then that is the way it functions for all of the atoms in the ship. Each atom has a force induced into it by the artificial time dilation which creates relative space-time for it. All of the atoms in the ship, cargo and crew have the same forced induced into them depending on their position within the field. The main goal is to envelop the ship in that time dilation bubble that is created to place the ship within a controlled relative space-time environment.

There may be multiple ways to simulate propulsion in space. Some may have to do with time dilation and relative space-time. Others may have to do with electromagnetism or mechanical manipulation of inherent energy and force or both. Possible ways to start experimenting could be the creation and manipulation of rotating magnetic fields.

Something to look further into would be gyroscopic procession and effects. The way rotational energy of a spinning gyroscope has procession affects when a downward force is applied to one side of its spinning axis. That would imply that the direction of force changes 90 degrees from the direction of applied force. That means if you apply a force in the direction of procession that will impart force 90 degrees from that which is upward. By adding more rotational speed to the direction of procession of a gyroscope based design the amount of lift would be proportional to that as well as the speed of the spinning flywheel. In that context it may be possible to create a seemingly weightlessness to be achieved.

This technique is technically not gravity or time dilation propulsion. It is simply the conversion of force of one type to another that may not have been fully investigated. There is always a loss of efficiency when converting forces in certain applications. This technique may not be subject to energy loss or minimal loss compared to the average application. This technique was introduced in the 1970's by Professor Eric Laithwaite a highly regarded engineer and visionary. As a result of his demonstration on live television a man by the name of Sandy Kidd built an experimental device that shows this concept and its very interesting and thought provoking to say the least.

A rotating wheel contains an inherent energy potential and that same concept may apply to an invisible rotating magnetic field in ways not currently understood. Since gravity is about energy potential then there is a twofold benefit of this method. The actual mass of the rotating mechanism has an inherent energy potential and the massless rotating electromagnetic field will also store an energy potential. Electromagnetic fields can contain much more energy potential and is only limited by the available electricity supplied as well as the mass of the mechanism needed to create the field and electricity.

Electromagnetic fields also release immense power and by collapsing a field that will release its power instantly creating super high voltage and current. A combination of these characteristics and factors may be involved to design such a device. In a spinning and rotating electromagnetic device if spinning fast enough time dilation also starts to alter the magnetic field. The faster the field spins the greater the effect time dilation has on that portion of it. In a rotating field the portion of the field with the greatest speed is the outer and farthest from the center portion. This method creates a situation where time is slower for part of the electromagnetic windings with the other part in faster time. The energy in the faster time winding (center region) may force the energy to pass through the slower time winding (outer region) creating additional relative effects.

***

The conception of propulsive gravity is to use the natural relationship of the nucleus and electron field of an atom to create propulsion. The idea is to recreate the potential energy in the atoms of the craft in the same way that time dilation effects space-time. By creating relative space-time in which the relative difference of time creates differences of spatial dimension. The difference of spatial dimension creates two sides of a field in atoms that differ in energy potential. For the atom that presents itself as the nucleus to be offset towards the opposite direction an object is moving under the influence of gravity. The craft must offset the nucleus in the opposite direction of desired travel. If you reproduce that effect then you create energy potential and propulsion.

The natural effect of relative space-time inversely places the electrons further and closer to the nucleus on opposite sides to create the propulsive force of gravity. The concept does not just apply to the nucleus/electron field it also applies to the field of the nucleons as well as the quarks. Any constituent of an atom that exists as a field may be induced with a higher energy potential by the relative space-time of time dilation.

***

In another example if a craft were to encompass itself with a magnetic field with the positive at the top and negative at the bottom it may create the same force of propulsion as the natural effect of gravity. The negative field of the craft would repel the electron field of the atoms of the craft upward and the positive portion of the field would attract the electron field of the craft upward. That would create a potential energy in the field of the atoms of the craft by altering the position of the nucleus in the same and similar way relative space-time does. The difference being the craft is what determines where that potential energy manifests. Once potential energy is manifested in the atoms of the craft that induces the source of movement aka propulsion.

That concept is based on the fact that the mass of an atom comes primarily from its nucleus with the electrons mass being essentially negligible and negated to zero because its .02% of the atoms mass. That means that the field of electrons can more easily be shifted in any direction to simulate an offset nucleus and therefore create energy potential in the atoms field. Shifting the electron field creates a distance differential between the two halves of the field and using Coulombs law creates a higher potential where the distance between the electrons and nucleus are closest.

I do not know if manipulating the position of the nucleus in an atom has any side effects on the passage of time. This process simulates the characteristics that time dilation has on atoms to create gravity as propulsion. I do not know if the process works in reverse and by simulating the effect you also simulate what causes the effect. It would be interesting to find out.

***

In the new ships using gravity propulsion you can increase the power without any negative effects or inertia on the cargo or crew in the ship. When a propulsive force is applied as potential energy in a field it applies a force to every part or atom of the ship and its contents. Unlike rockets of today that all of the effects and the forces the rocket engine is creating is felt by the ship, cargo and crew. When the rocket changes it trajectory the cargo and crew continue in the initial direction thus feeling the momentum or inertia being released on them. During lift-off the astronaut feels that energy being applied to him or her as increased g-forces. The ship also has to undergo the stress of that additional force of acceleration. When it gets into an orbit and no longer accelerating they do not feel the G-forces any longer.

What they do not feel is the amount of energy that is now stored in the atoms of their body or the difference in its fields in motion within the space-time field that constitutes the inherent energy. All of that energy applied to get them into orbit didn't just disappear when they became weightless. The mass of the people and the rocket and cargo all have stored energy as momentum. In order to get the rocket and contents back to the earth that energy has to go somewhere. It has to either be drained off or reversed. To drain it that is done by having a capsule with a heat shield to let the atmosphere slowly convert the momentum as velocity into heat on atmospheric re-entry. To reverse it the rocket will have to apply the same amount of energy to reverse course as it did to get to its current position. For every action there is an equal and opposite reaction. For which energy is conserved and makes it an inherent property when traveling through space.

Even though there is technically no stress on the ship or its contents using field energy potential as propulsion it doesn't negate the momentum stored in its mass as velocity. There is still the same amount of momentum stored as energy within the ship and crew if the ship were to impact another object. The new design of ship simply has a better way to apply energy as force on its contents to change direction as well as reversing that energy or absorbing it.

Because gravity is a force this process is a much more natural way to travel.

That explanation was meant to highlight the difference in the current form of travel to the future form of space travel. In the gravity propulsion ships they do not exert g-forces on the ship, contents or crew at all. That is the nature of this type of propulsion. This type of design applies a force to every atom inside of the relative space-time bubble. It creates a force on the ship and you in it with the ship moving with the force and you are also moving with the force. It's as though everything in the ship as well as the ship has its own separate propulsion unit acting in unison with all others. The ship is accelerating upward with its own propulsion while you are also accelerating up with your own propulsion. This is why you do not feel any g-forces on you. You may not even know that the ship just took off and if the ship makes a quick left turn you will not feel that either. Remember every atom is having a force induced into it. The propulsion of the ship is creating the illusion of naturally falling towards the direction the ship is moving.

This new generation of spaceship using artificially created gravity or manipulated energy potential would have a huge benefit when it comes to maneuverability and be able to apply propulsive force in any direction. It also has the ability to apply or induce a force within the atoms of the ship for any duration of time. That is a great leap over the current form of space rocket. The current rocket design burns its engines and runs out of its propulsion in minutes and usually does it pointed in one direction to obtain its desired trajectory. If it is off on its landing sequence on a trip to mars and flies by without getting captured by mar's influence then the ship and crew are lost without the ability to get back to earth. It would be a great leap in technology to be able to get in a ship, turn it on and fly to a destination. To be able to take the scenic route and turn left, right or stop and enjoy the view. Rockets of today and the way they function you cannot do that because the force as momentum would have to be reversed which requires equal energy to do so.

Here is an analogy that compares the time dilation bubble created around the ship to the natural time dilation of the earth. Imagine that you are sitting on the moon looking at the earth and below the earth is a spaceship with the top of the ship facing earth. Now imagine that the ship is moving towards the earth. Because the ship is using a gravity propulsion method you see the ship moving but don't know what is moving it.

This example shows that everything individually in the ship is moving under the force of the gravity propulsion. There is no distinguishing difference in this example of how the energy potential in its field that the ship is occupying was created. It may have been created by the ship artificially or created by the earths mass as natural time dilation. That is the beauty of this process because it's a totally natural phenomenon. Just because we don't have the machines to utilize this process now it doesn't mean we never will. Now that we can see more clearly we can endeavor to find a way.

***

We all know that any spaceship or design we currently use has a limited maximum speed. That speed is about 99% of the speed of light assuming it has enough fuel. Since using that type of ship would take us over three years to get to the nearest star system that would not be very practical in the long run. What we need to do now is start looking ahead to see possible ways to achieve faster than light speed travel. The work done here and now will help us get to that point sooner rather than later and also help us explore our own star system. We have to think in leaps and bounds over current technology.

The way to do that is to focus on devices that can alter space-time. If space-time can be altered then it can be manipulated to achieve a desired characteristic. Specifically one that creates opposing fields of time that creates a relativistic effect on a ship. Imagine a ship that can alter the space-time it occupies by creating a bubble of inverse space-time around it. The inverse space-time consists of a differential of space-time similar to how time dilation affects space-time.

In this case the bubble would be independent of normal space-time and produce an onion effect inside the bubble. That is where the layers of space-time in the center have the slower rate of time and increase towards the perimeter of the bubble to a level where the rate of time is faster than nominal. By decreasing time inside the ship and increasing it outside of the ship a relativistic travel speed is accomplished. If you decrease the rate of time inside by ten and outside by ten over the nominal rate you have a product of one hundred times the ships nominal speed. In the case of achieving faster than light travel relativity is the key.

You can travel faster than light because the speed of light is not relatively constant. The speed of light is a constant but it isn't relatively constant. Light emanating from a star is moving at the speed of light, at the maximum speed limit but its passing through the time dilation of the star. The time dilation has slower time at the surface and time is increasing in speed with distance. That means if you compared the speeds of light to one another relative to any specific point then at the surface of the star the light is traveling relatively slower than at a point farther from the surface. It is also going relatively faster even farther out and faster farther out if you compare all points at once.

If you could measure real time speeds that the time dilation creates you would see that the light is moving slower nearest to the star and faster farther away. The light isn't actually moving slower at the surface time is only moving slower. When you compare the light in slow time verses fast time it has a higher relative speed in the faster time. That shows the relativistic nature of the speed of light. If light speeds are relativistic naturally then why wouldn't you be able to create a machine to utilize a similar process artificially?

If light is moving around the universe at all different speeds relatively and the speed of light is a constant and the speed of time is constant then time dilation is the root of all relativity. We see the universe only as we imagine, we aren't seeing it as it actually exits in reality. We see it from one perspective but not from multiple perspectives. That seems to be the key here to go above and beyond and strive to take steps in learning more about the world around us. Changing your perspective and point of view is akin to exercising your mind.

Who could have imagined that after all of these years and the technological leaps in rocket design made only in the last twenty years that gravity is a natural propulsion process.

When we finally learn how gravity works we also find out that we can put that theory to good use. To use the same process that creates gravity to propel any machine designed to transport things, cargo or people. What we see in science fiction movies we may be close to turning into reality and into science nonfiction.

I know it's a little early to make that prediction, they are now more dreams than anything else and a dream is the blueprint to build from. If you can dream it you can do it, everything monumental that has occurred in history has started out as a dream. What is exciting about this now is since we know how propulsive gravity functions we can put it to use and move forward into the future. We can take steps to alter our destiny rather than let the currents of time take us on an uncontrolled ride. If we take control then we set the tone of the journey and create our own destiny instead of letting complacency dictate where we go and how we get there.

Artificial gravity propulsion *"The act of simulating the propulsive mechanism of gravity by sympathetic means in order to reproduce the same effect"*

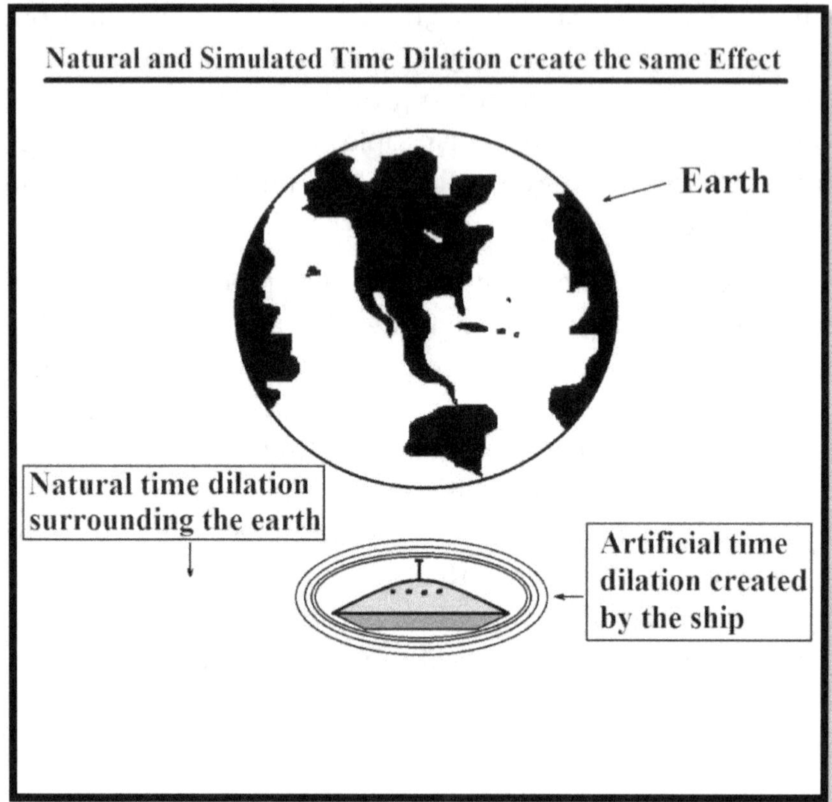

Illustration 5-1; Artificial verses natural time dilation

This shows an instance when there would be no difference in the time dilation between the natural created by earth or artificial created by the ship. The reason for this is because the earth's time dilation naturally causes objects to be propelled towards it. In this instance the ship is moving toward the earth, its time dilation propulsion would be in the same configuration as the natural time dilation surrounding the earth. The ship can increase or decrease the propulsion as needed to change the velocity or speed it is moving.

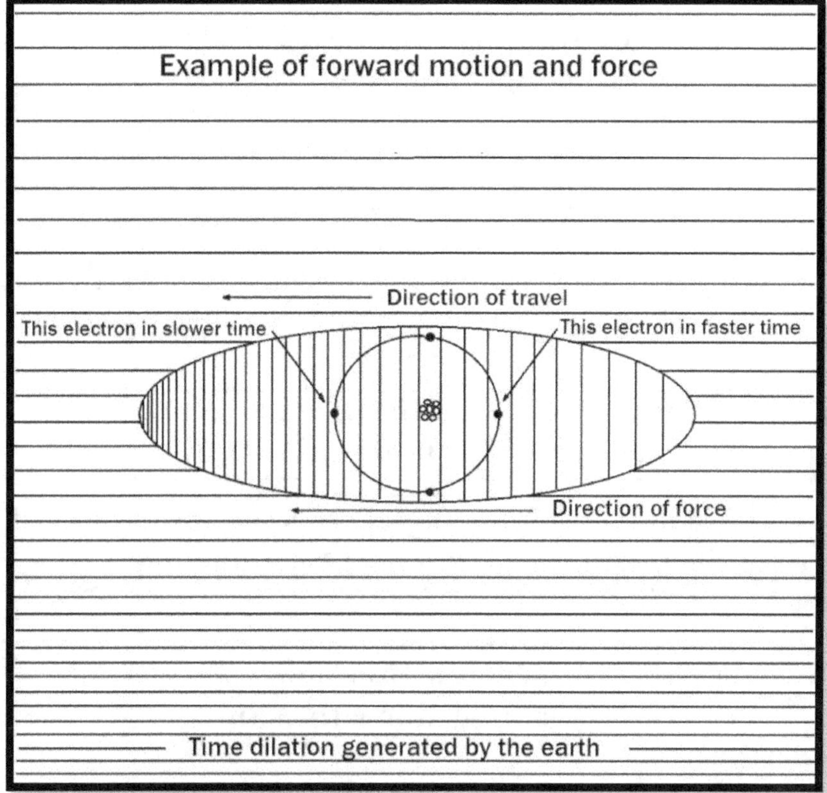

Illustration 5-2; Artificial time dilation bubble

This is the time dilation pocket created by a ship and showing how an atom in the ship would react to the direction of the time dilation. You see the main time dilation created by the earth and inside the pocket around the ship you see what the ship creates propelling it to the left. Essentially you see the atom falling towards the direction of force but since it's in a pocket artificially created its moving parallel to the earth. This means that if you can simulate time dilation it can be used as a means of propulsion.

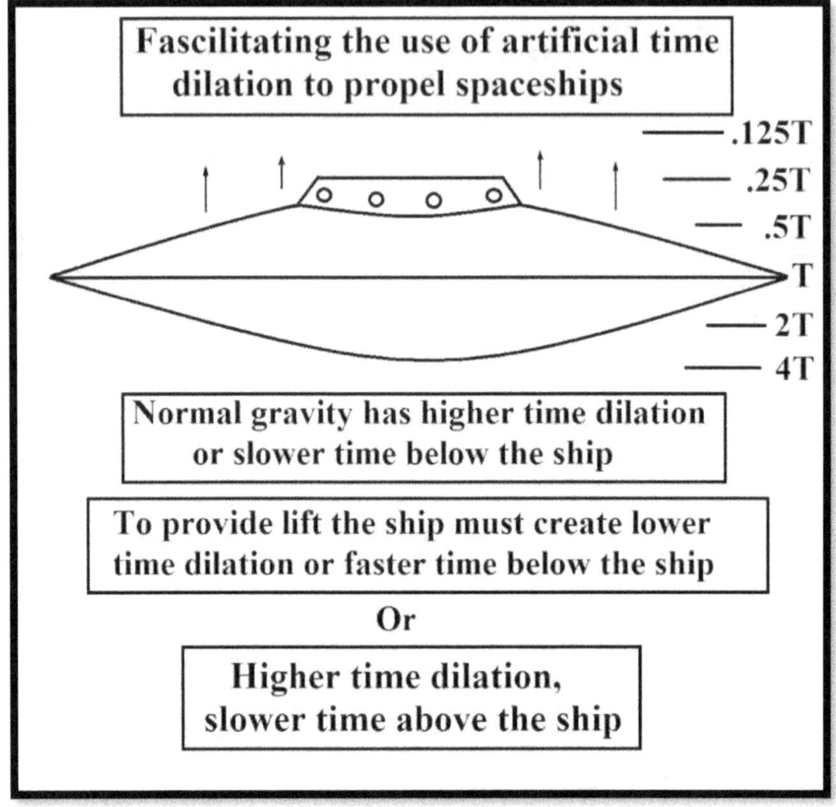

Illustration 5-3; Configuration of time dilation for lift

This is stating that the artificial time dilation it creates can be an increase or decrease, to slow or speed up time. If you slow time then that must be directed upward to move upward. Instead speeding up time below the ship would also lift the ship upwards. That is an example focusing on time only, an equal example could focus on spatial dimension because it's about relative space-time. That creates differences of spatial area which induce a potential energy used for propulsion like gravity.

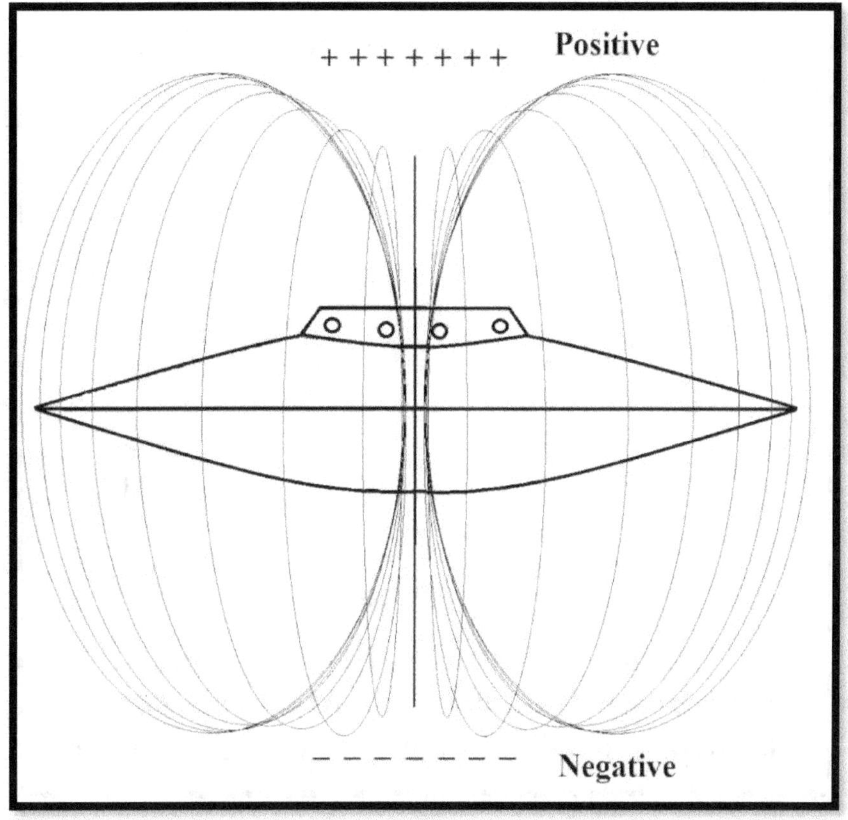

Illustration 5-4; An electromagnetic field to create potential

In this example if the ship can create a polarized electromagnetic field throughout it that may shift the electron orbit of the atoms of the ship. The negative polarity of the field would repel the negative electrons and shift them upward. That shift could artificially offset the nucleus and induce energy potential in every single atom of the ship. That would create the force to propel it. The next illustration is an example of one of the atoms in the ship.

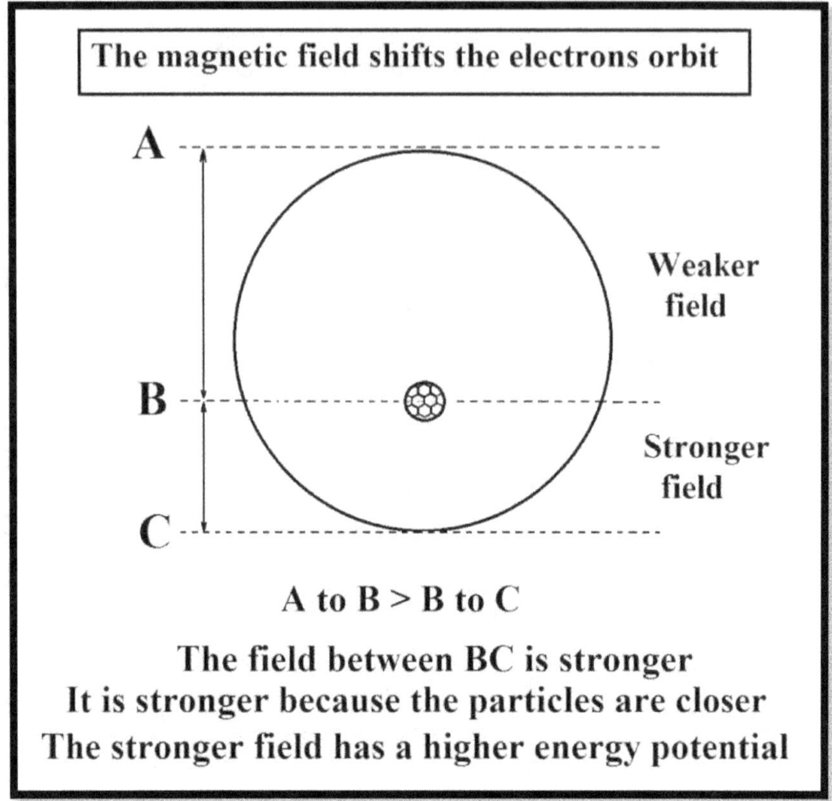

Illustration 5-5: An atom in a ship with an energy potential

This is an atom with an induced energy potential from the last illustration. The shift of the electrons causes an imbalance in the field which is the energy potential used to provide lift. Regardless of the process to create it this is and should be the end result to artificially create gravity propulsion. That means any successful endeavor to create this type of propulsion regardless if by artificial time dilation, magnetic field or other this illustration depicts the end result which is simply creating energy potential in the ships atoms.

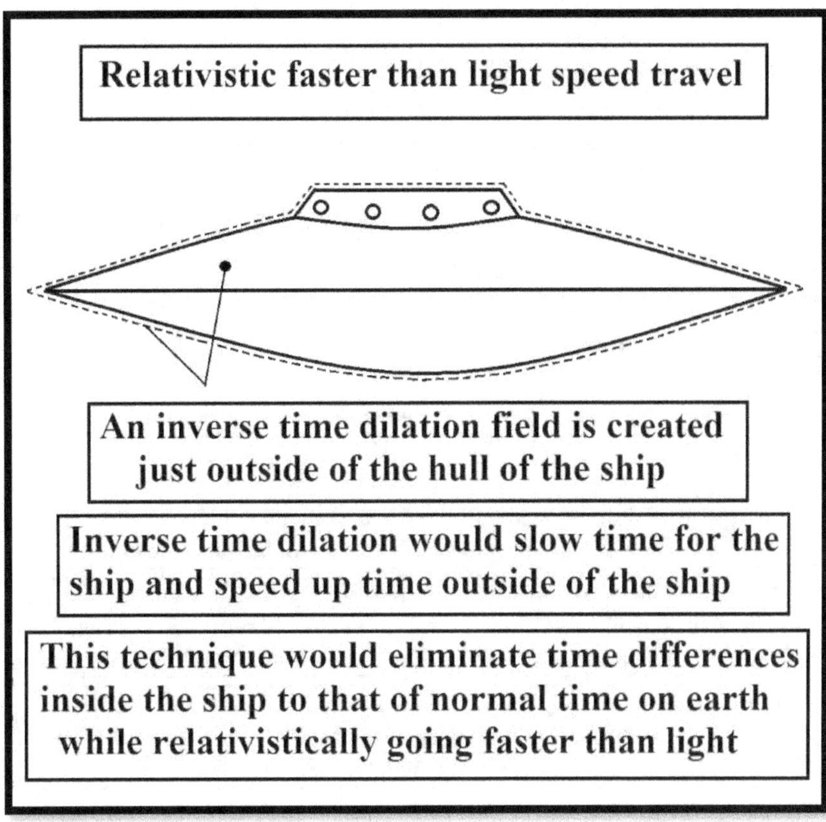

Illustration 5-6; Relativistic faster than light travel

To travel faster than light you must use relativistic methods. There must be an inverse time pocket around the ship so that the ship is in slow time and a pocket of space-time outside the ship is in faster time. This would multiply the ships nominal speed by the factor of the inverse time field. If you speed up time by 10 and slow down time by 10 in each direction of the inverse pocket the factor is 100. By using this relativistic method you can travel faster than the speed of light just the same as if it was a nominal speed. Time limits light speed, by manipulating time there is no limit.

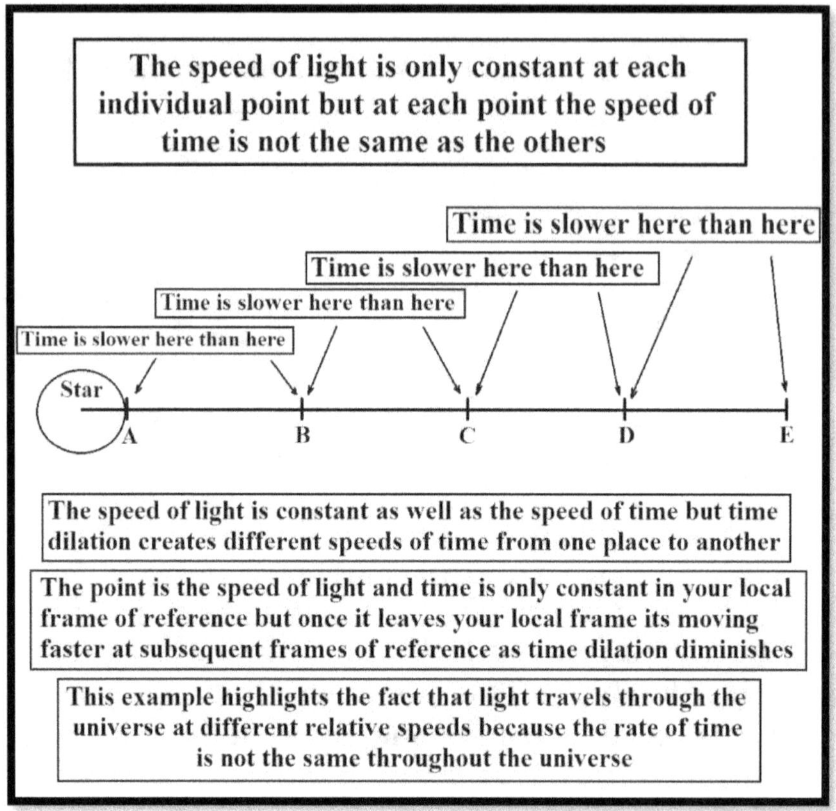

Illustration 5-7: Light is subject to relativistic differences

The speed of light has relativistic differences in speed when moving from an area of any time dilation outward. You could almost say that light from a star increases in speed the further away from the star it travels. When it's in slow or fast time it's moving at the speed of light but in faster time it's going relatively faster than when in the slower time. You could deduce that it increases its speed in a relativistic way as time speeds up. This also aids the thought process of having a larger spatial distance in slower time making it take longer.

A Link Between Energy Potential and Gravity as Force

During the time I was expanding on the scope of this theory of gravity and discovering that there was a field energy potential connection I began to contemplate if there was also a connection in other areas. If there was then the connection would strengthen this concept of gravity presented here. Since all energy and objects of mass and motion is really only energy in some form or another I wondered if the same basic principle would show up in other mechanisms. Whether there was a similar mechanism to the conservation of energy transferred from one atom to another. If one atom in motion impacting another that was stationary could be explained by the means of field energy potential transferring its energy from one to the other at impact.

I imagined an atom with velocity moving through space and impacting another that was stationary. The purpose of that thought experiment was to see if and how the electromagnetic field between the electrons and nucleus had any compatible similarities so as to strengthen the theory of gravity directly relating to energy potential in a field. The plan was to use this thought experiment to gauge the validity of this new gravity concept. To create a characteristic of interlocking puzzle pieces between what is known and what is proposed. So it could become self-evident and clearer in the mind's eye to all subjecting themselves to a new way of perceiving gravity. A new theory of gravity has an uphill battle and I wanted to do what I could to lessen the incline.

Gravity is so basic and elementary it didn't take long to find a connection between field energy potential and kinetic energy transfer in atoms. One of the biggest benefits of finding the connection is that it strengthens the characteristic that gravity is indeed a force. If it can be shown that gravity has similar characteristics within the scheme of force, momentum, potential and kinetic energy or are related in some way it is easier to group them together to link gravity and force in this context. Regardless of your philosophical bend gravity is the result of the development of potential energy and only depends on your view point as to how that is created. Looking at the broader picture of how the processes of energy potential manifests for objects of mass in the classical or Newtonian sense shows that the processes are all interconnected.

This new way of seeing gravity and force in objects of mass also sheds some light onto what defines mass and inertial mass. In my opinion which is based on conceiving of this process I tend to think mass is simply the result of how something is structured. A single electron or photon is said to have little or no mass. An atom has mass primarily in its nucleus which is constructed from protons and neutrons and even deeper a gluon field and quarks, basically multiple opposing particles or charges. So I say to myself that mass comes from the inherent design of being constructed as opposing charges or technically fields created by their opposing nature. If that line of reasoning holds true then it may put the pieces of the bigger picture together. It may show that gravity, force, momentum, potential and kinetic energy, the conservation of energy and all things related can be explained using descriptions of fields. Furthermore show how the stage or fabric of space-time is all that we need to express their mechanisms.

What we consider space-time is a field of its own with checks and balances. Apparently things cannot move through this field without energy applied or if moving cannot stop or change direction either. That is the inherent nature of space-time when looked at in the context of the law of the conservation of energy. A stationary object will remain in that state which seems to suggest that space-time is a neutral field but I suppose it cannot be termed a field if it isn't made up of opposite polarities though the opposing condition of space and time may be its "field". So maybe space-time shouldn't be called a field in the normal sense but can be considered to act like one. A better understanding may come from imagining space-time as a fabric where space and time have an inverse relationship in order to maintain a balance. It may not be a field in a normal sense of differences of charges but may always act to maintain balance to remain neutral, with concentrations of mass creating an imbalance and the inverse nature of space-time reacting to it.

Mass initiates time dilation from a governing mechanism of space-time or more specifically space-time creates time dilation to govern mass. The balancing effect of that slowing of time is the increase of spatial volume which is at the heart of altering the position of a nucleus in an atom due to the inverse square function of the effects diminishing of intensity.

With the model of gravity proposed in this book it is easy to apply that same process relating to mass, inertial mass, force and momentum. They are all connected and all use the same fundamental process with how they act and react. It is all about the opposing fields of the design of what we consider mass. What we consider to have the greatest mass is a combination of opposite polarity particles or charges. The electron has a negative polarity and the nucleus has a positive polarity in the design of an atom. One of the key points of the process is that the electron alone is considered to have a mass negated to zero since the mass of the nucleus is about two thousand times greater than that of its orbiting electrons. That point is of importance when showing how kinetic energy is transferred from one atom to another. It also shows one correlation between the inception of potential energy in this process and that of the mechanism of gravity.

The important factor for both processes is the placement of the nucleus in relation to its field of electrons. Not to exclude the nucleus because it may also be affected in the same way. By a higher concentration of nucleons in the faster rate of time which has the lesser spatial dimension. That also creates a higher energy potential in basically the same way the electron-nucleus field does but one view is all that is needed here.

It has not been conceived of previously that the nucleus of an atom is anywhere other than in a direct center placement of an atom. Though that thought should only be imagined for atoms in a zero gravity environment. An atom within a gravity field of relative space-time has its nucleus offset towards its trailing edge while it is moving towards the source of time dilation. The proximity of the nucleus closer to the trailing edge is what gives it a higher potential energy because the offset creates an imbalance of the atoms electromagnetic or electrostatic field. Inside of the nucleus there is also an offset that occurs and functions under the same principle but centers on the fundamental strong force. I know this is revisiting the process of gravity in a previous chapter but expressing it here in relation to the process of kinetic energy transfer may show it in a different light. This may process in your mind more clearly while you rationalize both processes at the same time.

***

The thought experiment is as follows so just imagine you are looking at an atom moving through space at a high velocity. You can see the nucleus in the center surrounded by a cloud of electrons orbiting or consisting of any theorized configuration. That atom is headed toward an impervious "theoretically solid" stationary wall, in this first thought exercise we will be looking at how the atom reacts to hitting the wall. Two things to remember are that the mass of the electrons in the atom are negated to zero and all of the mass is then technically contained in the nucleus.

The atom impacts the wall which then immediately stops the motion of the field of electrons since it technically has little to no inertial mass. The nucleus having mass, inertial mass and momentum continues to move but the electron field has stopped which sends the nucleus closer to the electron field of the impacted side. That creates an imbalance in the electromagnetic field of the atom and induces a higher energy potential in its impacted side due to the lesser distance between the nucleus and the electrons in that side pictured as an offset. The higher energy potential in the atom is pointed towards the direction from where it came. It then converts to kinetic energy or movement in that opposite direction and sends the atom back to where it came from. The kinetic energy inherent in the atom was transferred from one direction or vector to another solely by the change of the placement of the nucleus on account of its mass. The transfer mechanism was the conversion of kinetic energy to potential energy in its electrostatic field. The potential energy created the motion that propelled the atom back in the same direction as it came when it converted back to kinetic energy.

The placement of the nucleus reverts back to the center when the atom changes its direction after impact. The electron field re-centers itself back to equilibrium with the nucleus. The center position of the nucleus in an atom represents a fixed velocity in contrast to an atom that has the mechanism of gravity acting on it or a state of acceleration. That would represent a constant increase in velocity since the nucleus will be off center and produce a constant energy potential within its fields. An increase in velocity while in a gravity producing field represents a continued application of force as if the atom was propelled by a tiny field potential propulsion devise.

Let's look at that thought experiment using a different example, where an atom with velocity impacts a stationary atom with the same mass and transfers its kinetic energy. In both atoms the nucleus is centered, the impact scenario would be this. The electron field of each would be easily moveable due to their opposing negative force and minuscule inertial mass. At impact the electron field in motion would shift the electron field of the stationary atom. The nucleus with momentum would remain in motion as the electron fields' impact one another. That would do two things, send the nucleus in motion closer to its electron field and push the electron field of the stationary atom closer to its nucleus. That would create a higher potential energy in the side of impact in both atoms. The nucleus in both atoms would now be offset toward the side of impact creating potential energy in both atoms. That higher potential energy stops the atom in motion and intern creates the force to propel the stationary atom in the same trajectory and speed as the impacting atom because of the energy potential induced.

The energy is conserved when it transfers from one atom to another because there is no other mechanism to lose energy to. The one process using electromagnetic characteristics is at the heart of these examples and is mirrored in the mechanism that creates the force of gravity. In the context of this thought experiment the inside of the nucleus may follow the same characteristics it is just easier to explain using the electron/nucleus field.

These examples most likely also apply to any object with a force applied to create acceleration. In reality when you apply force to an object you are applying it to the electron fields of the atoms it consist of. All objects of mass undergo the same process that converts that force applied to potential and then to kinetic energy. The nucleus for which contains the mass is resisting the movement so therefore will remain offset until the acceleration subsides. At that point the electron field repositions back to balance ending the energy potential and the increase in kinetic energy level peaks.

An example is when a rocket takes off and accelerates the propulsion is really applying force to all of the electron fields in its atoms. The electron fields shift around the nucleuses causing them to be offset because it is resisting the acceleration due to their higher inertial mass. That process also correlates to an offset nucleus and energy potential in its fields. It is the mode of operation whereas the force you apply to something induces an energy potential in it by offsetting the nucleus. With that said it is also the mechanism of the reverse of that with any situation where a moving object releases its inherent energy with an impact. In that case the kinetic energy as momentum of the nucleus would offset and create potential energy and if there was no mechanism to conserve it would then create a force of impact.

It could be said that all energy applied to an object of mass regardless of the energy or force will manifest an energy potential by the means of offsetting the nucleus. The nucleus offset would be universal to adding to the inherent energy stored as momentum. It would apply to a broad spectrum of circumstances where a single characteristic applies across the board. Any and all examples can be defined using this characteristic. The offset nucleus would only be offset during acceleration, transfer of kinetic energy to potential, releasing its kinetic energy during impact and the initiation of kinetic energy as a result of gravitation. If you look at the full scope of what this chapter is proposing and consider it a most likely characteristic of how energy changes from one form to another then you can also use it as a gauge for this theory as well as for general relativity.

I mean no disrespect towards Albert Einstein and my admiration for him is not affected when I try to dispute some of general relativity as he basically created the foundation for this theory of propulsive gravity. I find his equivalence principles though in question when equating them as he did to gravity. One I find hard to accept as I've mentioned previously is the reasoning behind the explanation of gravity to an object at rest on the earth. This chapter's content that explains energy processes more specifically as energy potential highlights the inadequacies of the equivalence principle associated with it.

Even though gravity can be duplicated with acceleration of an object which seemingly links them together the mechanism between them is not totally the same. It is only similar by both having an energy potential manifest in objects of mass as a result of the offset nucleus. Relative space-time creates the offset nucleus for the process of gravity. Acceleration along with inertial mass of the nucleus creates the offset for that instance. Both processes create energy potential because they both create the offset nucleus and that is where their similarity lies. But the offset nucleus is 180 degrees opposite of one another.

For gravity the energy potential is manifested in the trailing side propelling an object towards the earth. In an accelerated object lifting up off of the earth the offset is also towards its trailing direction. If superimposing those two processes into one it is clear to see they are not the same and are physically opposite due to their opposing vectors. Equating acceleration and gravity to one another is incorrect in the context of this propulsive view of gravity and is not consistent with a standard view of energy potential. In GR it would suggest that kinetic energy in the object would be steadily increasing due to a constant energy potential being applied in an upward direction and that is not the case.

In reality acceleration only mimics gravity as a result of the inertial mass inherent in its nucleuses by the addition of force that creates the energy potential during the nucleus offset stage. It is clear that it is an actual process so creating an analogy of acceleration for why objects remain at rest on the earth is completely opposite and off base. Even though Einstein was correct in a way that acceleration gives the illusion of gravity that does not mean gravity is an illusion or of an abstract nature. This content may not have an effect on changing minds about how gravity functions due to the general consensus already established. But a general consensus is not what ends up giving us facts physics does.

***

The main reason for this chapter was to show a correlation between certain aspects of the process of propulsive gravity focusing on electromagnetism that transfers kinetic energy from one object to another. In that process kinetic energy converts to potential energy at the moment of impact nullifying and transferring its energy to the other atom. It induces a potential energy within both atoms fields which stops the atom in motion and becomes kinetic energy in the stationary atom.

The center of this process is the point kinetic energy is transferred atom to atom as field potential energy. The mechanism for this force exchanging atoms is the same as for the theory of gravity being described in this book. Simply potential energy is created which induces motion aka kinetic energy. Force is associated with this mechanism but is not so easy to see when looking at the beginning of the process but is undeniable when an apple hits you on the head.

Field potential in different phenomenon *"Multiple actions as a result of field energy potential that share a base factor between them in creating, sharing or transfer of energy"*

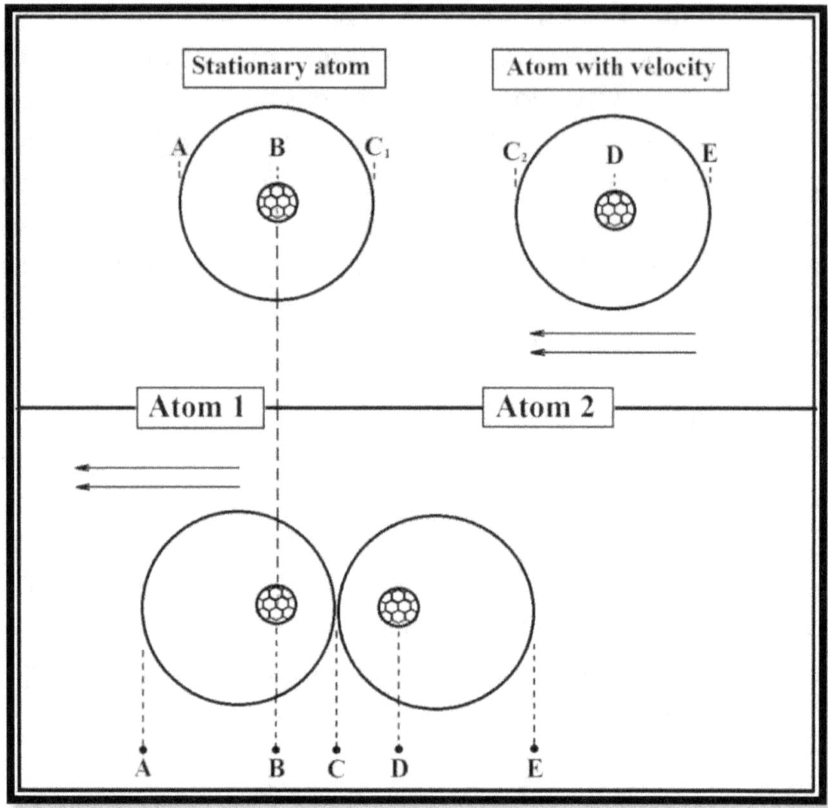

Illustration 6-1; Kinetic to potential to kinetic energy

An atom in motion that has kinetic energy transfers that to potential energy then back to kinetic energy in the impacted stationary atom. At impact atom 2 nucleus's momentum sends it closer to its electrons forcing the electrons of atom 1 to shift around its nucleus. The offset of nucleus's manifest as potential energy and transfer from atom 2 to atom 1. When all the kinetic energy has transferred from atom 2 atom 1 will contain it and be propelled. As it moves both atoms electrons will reposition to center around the nucleus.

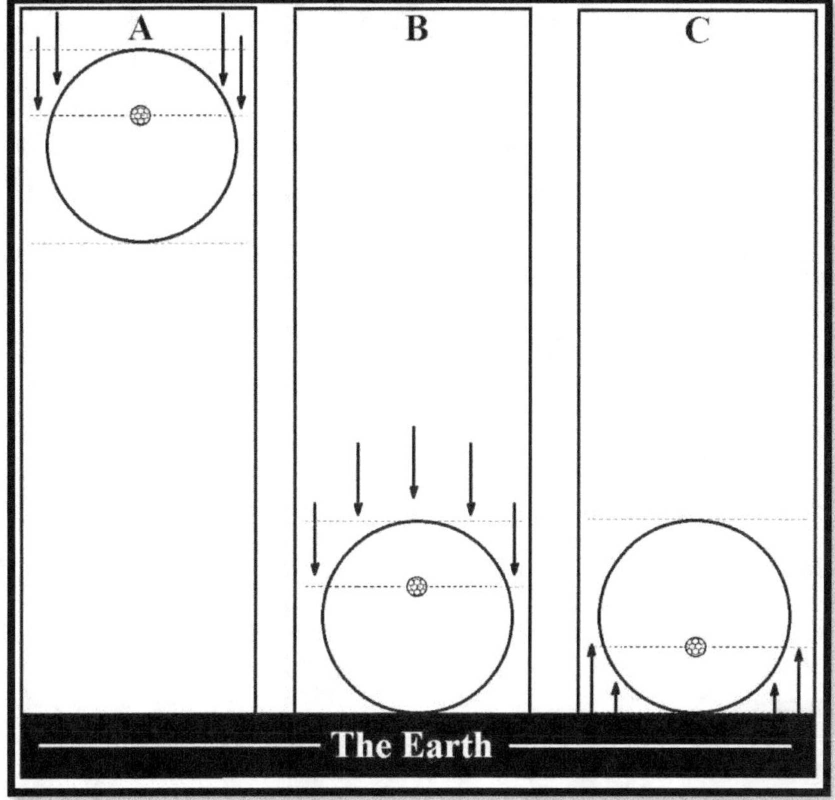

Illustration 6-2; Nucleus offset field potential comparison

A and B show the nucleus offset that represents energy potential in the atoms field to accelerate down and be held to the earth. C represents general relativities assumption that the earth is accelerating up and shows where the nucleus offset would be if that were correct. This example shows how the equivalence principle for which C is based on is the complete opposite of the field energy potential view given in this book. Besides that point if there was an acceleration happening in C upward then that atom would have induced in it an increasing kinetic energy and momentum.

Conclusion

The goal of this book and theory of gravity presented is to create a better understanding as to how the mechanism of gravity functions. Up until this point it is my belief that how we define gravity has remained incomplete. That is understandable because that is the journey that the process entails in going from unknowing to knowing. In completing the journey from page one to the end in this book you were presented with an opportunity to move further into the knowing. As society assimilates these concepts it too becomes more enlightened with a knowledge that may help us progress out into the star system we call home and see what is out there with a curious awe. That thought alone invokes an enthusiasm that we all have inherently within us to go out among the stars to explore.

My plan was to describe my thoughts and theories clearly and simply to let the information and concept speak for itself. To put my theory out there in print and let the reader be the judge just as Isaac Newton did at the end of his manuscript on gravity. The concept presented here is more of a completion or complimentary to Newton's findings on gravity because they agree for the most part and explain the mechanism he left up to others. I think that the changing of the perception that gravity is propulsion rather than an attraction leads this theory toward a better understanding since he covers gravity fairly thoroughly except for the final mechanism.

It is not misunderstood that Albert Einstein's general relativity plays a large part in the base of this theory or concept. Without the factors of time dilation and relative space-time this concept would never have materialized. Without first pondering time dilation and the relative effects on space-time my mind would have never been able to link the puzzle pieces that constitute this concept into the steps that immortalize it. Yes there are some differences with general relativity just as there are with Newton's gravitation but are more fundamentally opposing. So in this instance I would not suggest it is complimentary to GR but more of a redefinition, a general relativity 2.0 to through a term out there.

It was said that Einstein's general relativity overtook and replaced Newton's theory of gravity but in my opinion now after the fact simply supported and strengthened it. Einstein took gravity to a new level by theorizing the factor of time dilation and relative space-time that is true but didn't go far enough to cement the status of GR as a whole to be the final explanation of gravity. As the writer and creator of this theory of gravity it may not be bold to say it combines Newton's and Einstein's views of gravity into one.

A big assumption with general relativity is that gravity in its context affects the trajectory of light resulting in it bending. Light does bend due to mass in a sense that is true and is proven but what is not actually proven is how. What goes hand in hand with mass occupying space-time is the effect of time dilation and that initiates the bending of light. It is the same with gravity which is the result of relative space-time. The relative space-time is what bends light and just because it shares that with gravity it should not be misconstrued that gravity bends light.

A similar situation applies to gravitational waves for which do exist but is not produced by gravity itself. What are called gravitational waves are really only waves of time dilation which exist as waves of relative space-time. Again if something has a similarity that does not mean it has the same mechanism as gravity. A moving wave of relative space-time does not have the same effect as a stationary constant source of relative space-time created by time dilation. A moving wave would simply come and go and the effects of the relative space-time would cancel as the wave passes. It is the same as the effects of time dilation that light or energy produces, if it's moving it produces a wave in space-time which does not produce the mechanism of gravity. Because of this gravitational waves do not exist technically because the terminology is inaccurate.

The explanation of GR on the mechanism that holds an object to the earth is another thing to look at that places doubt on it. Since Einstein rationalizes that the effect of gravity is the same as a body being accelerated he mistakenly put too much relevance on that in creating GR and that was a misstep and assumption. To say gravity and acceleration on a body are the same thing or create the same effect in the same way lead him down a rabbit hole to an inadequate conclusion. Looking at how that may be true in space is at least conceivable but the same analogy doesn't hold water trying to explain objects at rest on earth. To say you are being held firmly to the planet because it is accelerating upward through space-time is in fact GR's weak link. Most people with an inquiring mind will surmise that since a person on the opposite side of the planet was under the same condition would wonder then how that could be the mechanism. Placing time as a fourth dimension to try and explain it by saying we are moving through time doesn't clarify things at all.

These conclusions associated with GR in my opinion define it as being abstract. Another issue I have with it is that gravity is considered as not being a force. Theoretically speaking in the context of a mathematical equation if force were to be on one side would it then not be on the opposing side in some form or another. Math may be an abstract representation of what is real but solutions are designed to have real connotations. I cannot conceive of a meteor impact and the force released not coming from something initiating it that was not a force. The force of impact "was" initiated as energy potential which became kinetic energy and force is a product of energy.

***

Conclusion

The path we followed that got us here to this point in time is one of complacency. Society has become complacent in defining the world around us with accuracy. It became enamored with bigger and better explanations just as advertisers do to sell a product. We all know that life is one big maze and navigating it is hard enough, we don't need any influence pushing us down the wrong path. Creating our own journey is a drive that keeps us motivated even though we don't know the true passage that solves the maze. It is the hope and faith that the completion of life's maze leaves us with a view of the bigger picture which feeds our spirit to keep moving and not fall behind.

I believe it is inherent in us to look toward the stars, the wonder and amazement of what could be out there is undeniable. If that were not true we would be content on staying put on the earth and burying our heads in the sand. Maybe it's the result of not wanting to be alone in the universe and to search for neighbors to satisfy a social characteristic. To also search for a greater sense of knowing of what is out there and how does the universe function. We are still at the smallest point and searching for answers to the largest questions.

Being held back on our own accord has done little to progress space ship related technology. In over a hundred years we are still at the point of going to space in machines that are equivalent to large controlled explosive devices. The propulsion for those devices lasts only minutes, I think we can consider those to be first generation devices. The question I ask is when are we going to start developing second generation devices? Space ships that create the power to propel them right on board with the ability to turn it on or off at any given point. To have the ability to create a method of propulsion that is unidirectional and more versatile than current methods.

The ability to do those things is at hand because now we have a view of the mechanism inherent in the process that creates gravity to start from. Gravity is natural propulsion and it has been in front of us all along. I hope this sends sparks of imagination through everyone's mind to initiate the conception of new technologies. The one hundred year long drought has subsided and we now have a better view of the mechanism of gravity and the next step is to utilize it. Going to mars is interesting to imagine but first you must use your imagination to turn that into reality.

I believe that since we are now more aware of how to get over the hurdles we can do so without tripping and falling on our face. It gives a place to start because we know where we want to end up. Use the "philosophy" within this book as a blueprint and read it with your own unique way of comprehension. A journey of a thousand miles begins with a single step. That is the real reason having an accurate theory of gravity is important.

I didn't write this book for some whimsical flimsy run of the mill theory. I wrote it because I am totally confident in it to be justifiable and accurate. I want it to be the first step in helping us achieve a new level of technological advancement. I wrote this book to provide an accurate common sense explanation for the process of gravity and by coincidence also developed the means to start developing a ground breaking revolution in transportation and space travel.

Conclusion

This started out as a book on a rational, analytical common sense theory of gravity. It was spawned not from science fiction or far outlandish ideas but of pieces and characteristics of this universe that already exist as fact. It began with Einstein's theories of relativity and time dilation. Also Newton's tried and true theory of gravity that provides the biggest picture for which to start from. Coupled with the knowledge of the atom and my own interpretation of the relationship between perspective or relative space and time we get a better view of gravity.

We find out that gravity isn't just something we experience and have to work against to get out there into space. Gravity as a mechanism that we can use to utilize its characteristics in order to simulate a ship falling towards the earth but in reality sending it out to mars. It's an honor for me for you to have read this book and to share in the dream. It's a dream to see the stars up close, to be out there in space to experience the adventure and the unknown and find new habitable planets. You have just taken the first step to make a little bit of that unknown known. We are at the dawn of the next greatest adventure of a lifetime, buckle your seatbelts.

About the Author

The author Scott Calvert is an American of Norwegian, English, German and French decent. His father's heritage is Norwegian and English and his mother's is German and French.

He grew up following the standard accepted version used to describe the universe and gravity but that wasn't enough for it to take hold of his mind because of the irrational and inconsistent explanations. He resisted those because of the need for a rational and common sense view. That sowed in him the ability to see things somewhat differently which lead him to his own ideas and theories instead of following others and accepting things simply because there is no other explanation being provided.

His fascination with the universe was sowed at an early age with the description that the universe was infinite in size and that being a paradox puzzled him which would occupy his mind throughout his life. After spending most of his lifetime working out the details of the universe he is finally putting those thoughts into these two books. They are the end result of him following his fascination of the subject matter. He is aware of the tall mountain he has to try and climb for these books to be accepted. Writing these books to try and change the general consensus of cosmology was not a choice it was seemingly a directive from the universe. After the answers to long sought after questions were conceived he was compelled to share them.

He published this book to give other people the ability to see that there are alternatives to commonly excepted theories about what is normally accepted and not questioned. To get away from old and antiquated theories that do not make rational sense and replace them with new common sense versions that provide a broader and more accurate picture.

This is his second book on the topic of theoretical physics with the first covering the entire sphere of the universe. His first book shares his unique vision of the creation of the universe and the cycle that it follows. The dissection of black-holes from beginning to end and the part they play in the transition of the universe from one incarnation to another. He describes how he believes time dilation is the mechanism that produces the effects that we see that suggests dark matter theory should be accepted at face value when indeed it should not. Since he isn't tied down by preconceived ideas and dogma he thinks for himself and looks at things from his unique perspective to really get into how things work and function. The universe is the most exciting and thought provoking puzzle in his lifetime and he enjoyed disassembling and then reassembling it.

***

I was born in the mid 1960's and like any child my spare time was designed for playing. I would play with building blocks, tinker toys, Lincoln logs or Legos as well as many other toys that would exercise my imagination. It seems building things was a common denominator when it came to which toys I would choose because that would allow me to exercise my creativity. As I got older that transitioned to model cars, airplanes and rockets. Science related projects took hold as this was only a few years after the Apollo moon landings. My older brother had some influence on science related things I was introduced to since that was his interest otherwise I may have never gone in that direction.

Getting back to 1969 and the Apollo moon landing, I was still a child at the time and like I mentioned my life was devoted to playing. I recall only vaguely seeing some news footage of the event but was not overly interested. I wish I had been, it occurred within my lifespan and it was a pivotal moment in human history. I wish I had paid more attention to it then because the endeavor now to go back to the moon and go to mars brings back those youthful memories and makes clear what I missed out on. I feel excited to watch the Spacex Starship prototype test flights and I think now as an adult I can appreciate the experience more than I would have as a child.

Way back then I may not have consciously taken interest in the Apollo moon missions but it seems that somehow my subconscious mind must have been affected by it. I suppose that's why I've always favored science fiction space related movies and shows like Star Trek, Star Wars, Stargate, Farscape and Dr. Who to name a few. I don't know if you have to learn to be fascinated by these types of shows or have them be a part of your youth or if science fiction is simply in your blood. It's difficult to explain where it comes from but I suppose it's simply there as part of your brain function as a result of how it was developed in your youth. Dreams as a child become your reality as an adult. When I was a child I dreamed of building an anti-gravity spaceship. I would doddle in a book and make drawings of what I imagined a design would be, I'm sure they have no bearing on a real design as I was never really satisfied with them and even as a child I knew they were only basic doodling's.

It's a strange thing though, I made those drawings as a child about fifty years ago and now I'm working on a book focusing on a theory of gravity as propulsion. I think to myself will I actually figure out how to build an "anti-gravity" spaceship in my lifetime? I know it's not anti-gravity but that's what my childhood mind was trying to achieve. My childhood dream is now closer than it's ever been to becoming a reality since it has taken a new form by developing a concept to make it achievable. It seems like it may be part of an unknown plan and I'm headed in the direction that it may come full circle. I don't know if I will ever get there but it seems possible for that dream to come to fruition, it seemed like a pipe dream so long ago.